"Michel Landaiche has done us all a gr[eat service with this excel]lent writing on living and learning thro[ugh groups. As is typical of] his work, Landaiche brings his unders[tanding, scholarship, and] experience of working with groups. O[ver the course of writing abou]t being in group processes, Landaiche expertly navig[ates the complexity of grou]p life that, in its unfolding, demonstrates that neither he, nor us as readers, can be anything other than immersed in the business of being together. Transactional analysis had its early history rooted in group work, which whilst innovative in its time, has at times, been overlooked as the central arena in which our practice and training takes place. In this much needed title, Landaiche resets the dial, reminding us of the enduring complexity and promise of what it is to be with others."

    Giles Barrow, MA, MEd, PGCE, TSTA-Education, author and
    co-editor of *Educational Transactional Analysis*

"Perhaps the best treatise on the nature of collective life that I have ever read. Landaiche brilliantly integrates object relations and other psychoanalytic theories of the mind with transactional analysis, family systems theory and the author's own unique perspectives honed over the course of a lifetime in and out of the treatment room. Indeed, it is the interweaving of personal and theoretical that makes this work unique and compelling, providing the reader with both an overview and deeply sophisticated understanding of the complex nature of groups. This book will be of great value to anyone with an interest in how we learn, lead and function in groups—psychotherapy, classroom, societal and the gamut."

    Steven Kuchuck, DSW, president of IARPP,
    the International Association for Relational Psychoanalysis
    and Psychotherapy; editor of the Gradiva Award-winning
    *Clinical Implications of the Psychoanalyst's Life Experience:
    When the Personal Becomes Professional*

"Michel Landaiche draws up the work of the 'three B's'—Berne, Bion, and Bowen—to offer an understanding of group cultures, group process, and the experience of learning in groups. Written in a voice both clear and humble, this book addresses the somatic, psychological, social, and spiritual aspects of group life and work. While the content is rich and stimulating, I found the ways in which Landaiche involves himself deeply and honestly into his own learning in and about groups, an exceptional demonstration of compassion and respect for the other. I hope this book will be translated into French, so that I will be able to recommend it to my students and clients."

Isabelle Crespelle, psychologist; psychotherapist; teaching and supervising Transactional Analyst-Psychotherapy; co-founder of the French School of TA; and co-founder and former VP of the French Federation for Psychoanalysis and Psychotherapy

"This is a timely compilation of thought and practice about groups and social systems. The author brings his wealth of experience and reflection on the many aspects of conscious and unconscious group process. A must-read for organizational consultants and for any professional who works with the notion that the direction for organizational transformation is already existing in the client organization, only covered by unconscious inhibitions."

Servaas van Beekum, drs, TSTA-Organizations; winner of the 2015 Eric Berne Memorial Award

GROUPS IN TRANSACTIONAL ANALYSIS, OBJECT RELATIONS, AND FAMILY SYSTEMS

Groups are arguably an essential and unavoidable part of our human lives—whether we are a part of families, work teams, therapy groups, organizational systems, social clubs, or larger communities. In *Groups in Transactional Analysis, Object Relations, and Family Systems: Studying Ourselves in Collective Life*, N. Michel Landaiche, III addresses the intense feelings and unexamined beliefs that exist in relation to groups, and explores how to enhance learning, development, and growth within them.

Landaiche's multidisciplinary perspective is grounded in the traditions of Eric Berne's transactional analysis, Wilfred Bion's group-as-a-whole model, and Murray Bowen's family systems theory. The book presents a practice of studying ourselves in collective life that utilizes a naturalistic method of observation, analysis of experiential data, and hypothesis formation, all of which are subject to further revision as we gather more data from our lived experiences. Drawing from his extensive professional experience of group work in a range of contexts, Landaiche deftly explores topics including group culture, social pain, learning, and language, and presents key principles which enhance and facilitate learning in groups.

With a style that is both deeply personal and theoretically grounded in a diverse range of studies, *Groups in Transactional Analysis, Object Relations, and Family Systems* presents a contemporary assessment of how we operate collectively, and how modern life has changed our outlook. It will be essential reading for transactional analysts in practice and in training, as well as other professionals working with groups. It will also be of value to academics and students of psychology, psychotherapy, and group dynamics, and anyone seeking to understand their role within a group.

N. Michel Landaiche, III, Ph.D., has practiced for 30 years as a psychotherapist and group facilitator. He also provides training and supervision for counselors, therapists, and organizational consultants in his hometown, Pittsburgh, Pennsylvania, USA, and in Europe.

INNOVATIONS IN TRANSACTIONAL ANALYSIS:
THEORY AND PRACTICE
Series Editor: William F. Cornell

This book series is founded on the principle of the importance of open discussion, debate, critique, experimentation, and the integration of other models in fostering innovation in all the arenas of transactional analytic theory and practice: psychotherapy, counseling, education, organizational development, health care, and coaching. It will be a home for the work of established authors and new voices.

https://www.routledge.com/Innovations-in-Transactional-Analysis-Theory-and-Practice/book-series/INNTA

Titles in the series:

TRANSACTIONAL ANALYSIS OF SCHIZOPHRENIA
The Naked Self
Zefiro Mellacqua

GROUPS IN TRANSACTIONAL ANALYSIS, OBJECT RELATIONS, AND FAMILY SYSTEMS
Studying Ourselves In Collective Life
N. Michel Landaiche, III

CONTEXTUAL TRANSACTIONAL ANALYSIS
The Inseparability Of Self And World
James M. Sedgwick

GROUPS IN TRANSACTIONAL ANALYSIS, OBJECT RELATIONS, AND FAMILY SYSTEMS

Studying Ourselves in Collective Life

N. Michel Landaiche, III

Routledge
Taylor & Francis Group
LONDON AND NEW YORK

First published 2021
by Routledge
2 Park Square, Milton Park, Abingdon, Oxon OX14 4RN

and by Routledge
52 Vanderbilt Avenue, New York, NY 10017

Routledge is an imprint of the Taylor & Francis Group, an informa business

© 2021 N. Michel Landaiche, III

The right of N. Michel Landaiche, III to be identified as author of this work has been asserted by him in accordance with sections 77 and 78 of the Copyright, Designs and Patents Act 1988.

All rights reserved. No part of this book may be reprinted or reproduced or utilised in any form or by any electronic, mechanical, or other means, now known or hereafter invented, including photocopying and recording, or in any information storage or retrieval system, without permission in writing from the publishers.

Trademark notice: Product or corporate names may be trademarks or registered trademarks, and are used only for identification and explanation without intent to infringe.

British Library Cataloguing-in-Publication Data
A catalogue record for this book is available from the British Library

Library of Congress Cataloging-in-Publication Data
A catalog record has been requested for this book

ISBN: 978-0-367-36920-0 (hbk)
ISBN: 978-0-367-36921-7 (pbk)
ISBN: 978-0-429-35191-4 (ebk)

Typeset in Times
by Swales & Willis, Exeter, Devon, UK

CONTENTS

	Acknowledgments	viii
	Permissions	x
	Introduction	1
1	Engaged research	9
2	The shared bodymind	23
3	Learning and hating in groups	38
4	Social pain dynamics in human relations	55
5	Looking for trouble in professional development groups	69
6	Maturing as a community effort	88
7	Groups that learn and groups that don't	94
8	The learning community	113
9	Principles and practices of group work	123
10	Closing reflections	143
	References	*150*
	Index	*164*

ACKNOWLEDGMENTS

I have so many friends, family, colleagues, and mentors to thank for my maturing, my coming to understand the human condition, and my stumbling with determination toward a meaningful life, in particular my mother, Carol Ann Comstock Landaiche, and my father, Nemour Michel Landaiche, Jr.

In the context of this book, I especially want to acknowledge and express gratitude to my first formal group teachers and mentors, Nick Hanna and Vicky Lahey, both of whom brought their group sensitivities, enthusiasms, and skills to my early learning while I was in graduate school at Duquesne University pursuing my master's in counseling with an emphasis on group work.

As a subsequent step along this path, I was so very appreciative of the chance to learn the Tavistock group-as-a-whole method through the steady and powerful presence of Frances Bonds-White, who facilitated an experiential study group of which I was a member for several years in Pittsburgh. Her manner of being in a group still accompanies me through my more difficult encounters, offering me a way through to some emerging insight.

My understanding of group, family, organizational, and social processes was also greatly enriched by my encounter and subsequent long years of learning with Jim Smith, who introduced me to Bowen's family systems theory and the human science of phenomenology. As one of six cofounders of the Western Pennsylvania Family Center, Jim worked closely with and learned from principal cofounder Paulina McCullough. He also played an integral role in that organization's ongoing efforts to teach and continually learn Bowen theory, the framework of which provides the foundation for my theorizing and practice today.

Learning object relations theory was significantly mediated for me through my regular encounters, over a number of years, with Charelle Samuels, a senior psychotherapist and supervising consultant in my professional community. She offered a weekly reading group for many years along with her more embodied approach to psychodynamic theories,

ACKNOWLEDGMENTS

a unique perspective that integrated well for me with the group-as-a-whole methodology and the theory of natural family systems.

My subsequent thinking about social pain was greatly influenced by the work and writings of Herb Thomas, a psychoanalyst who also practiced in the Pennsylvania prison system. His views on what he called "the shame response to rejection" helped me identify, understand, and begin to work more productively with the often explosive reactivity that can occur in families, groups, and other collectives when the cycles of rejecting and counter-rejecting escalate past the point where any of us are able to think and stop the harm we are inflicting and receiving.

Finally, I have been indebted to the international transactional analysis community, with its long and solid tradition of exploring group and organizational processes as well as its attention to human learning. I have been especially influenced in my understanding of human development by the approaches and writings of Giles Barrow and Trudi Newton and by my opportunity to co-facilitate experiential learning groups over the past 15 years with William Cornell. His encouragement to me to write has also played an invaluable part in the life of this book, as has Jo Stuthridge's editorial guidance. It has also been a privilege to work closely with the ten dedicated members of the Romanian TA community: Traian Bossenmayer, Marina Brunke, Ioana Cupsa, Dana Anca David, Diana Deaconu, Irina Filipache, Radu Gheorghe, Karina Heiligers, Ioana Maria Pirvu, and Georgiana Rosculet. All of them have been part of my "Studying Ourselves in Collective Life" group, which has met annually in Bucharest since 2015. Finally, I am deeply grateful to Robin Fryer—the longtime managing editor of the *Transactional Analysis Journal*—who has served as a mentor for my professional writing. She has offered me a structure governed less by strict rules and more by the discipline and emerging recognition of the feel for clarity, which has also left me with sufficient space for passionate inquiry into the nature of my professional calling.

PERMISSIONS

The following articles were originally published in the *Transactional Analysis Journal*, © International Transactional Analysis Association, Inc.

Chapter 1 first appeared as "Engaged Research: Encountering a Transactional Analysis Training Group Through Bion's Concept of Containing," *Transactional Analysis Journal, 35*, 2005, 147–156.

Chapter 2 first appeared as part of "Skepticism and Compassion in Human Relations Work," *Transactional Analysis Journal, 37*, 2007, 17–31.

Chapter 3 first appeared as "Learning and Hating in Groups," *Transactional Analysis Journal, 42*, 2012, 186–198.

Chapter 4 first appeared as "Understanding Social Pain Dynamics in Human Relations," *Transactional Analysis Journal, 39*, 2009, 229–238.

Chapter 5 first appeared as "Looking for Trouble in Groups Developing the Professional's Capacity," *Transactional Analysis Journal, 43*, 2013, 296–310.

Chapter 6 first appeared as "Maturing as a Community Effort: A Discussion of Dalal's and Samuels's Perspectives on Groups and Individuals," *Transactional Analysis Journal, 46*, 2016, 116–120.

INTRODUCTION

Most of us, like it or not, are significantly involved with human groups of various kinds: families, organizations, communities, nations, and other human collectives. For better and for worse, our lives are socially entangled, whether we are designated as our groups' leaders or members.

To this day—and for all the years and effort I have put into studying groups and figuring out how to work with them—part of me still wants nothing to do with them. Part of me still believes that, one day, I can finally break free of collective life, that I can just get away from all these other annoying, sometimes frightening people!

Yet I have also observed, over and again, that every time I indulge these fantasies, every time I try to eliminate my part in community—in family, in organizations—I kill a part of myself. I actually only eliminate my chances of fulfillment, for which I need this fact and facet of my human life, rough as that fact can be, at times, to stomach.

Now, as I enter what may be, in all probability and with good fortune, the last 20 years of my life, I want to bring my group work to some greater integration, to some better understanding of my evolving outlook. I want to gather, revisit, and reflect on the meanings of this work largely to discover how it might help me in these final years to achieve what still matters greatly to me in terms of my own maturing; the welfare of my children, nieces, and nephews; and the growth of those who ask me to work with them to achieve their own maturational aspirations. I also want to collect these efforts for the benefit of my many colleagues and other fellow humans who are likewise involved in figuring out how to collaborate for the benefit of our collectives and the individuals who comprise them.

What can we learn together that might reduce the harm and suffering we seem unable to avoid entirely? Whatever our specific roles—as leaders or as members—how might we each want to show up differently, more productively, and with greater satisfaction within this key dimension of our lives?

I invite you to join me in this effort, as I felt welcomed by so many who were working on this for long years before me. In consequence of that transgenerational process, I stand in considerable gratitude for the benefits that have flowed my way. I look forward to learning what good may also flow toward, through, and from you, my readers.

Our personal histories with groups

What are the many ways we have each been affected by our lives within our various groups? How might we carry that history into our current lives, perhaps with little awareness?

For myself, although my formal interest in this topic began just over 30 years ago—when I went to graduate school to study group therapy, human dynamics, organizational processes, and family systems—it would be more accurate to say that groups have been massively compelling to me for well over 60 years. This began when I was born into the overfunctioning, eldest position of what would eventually become nine siblings. As a boy, I received early, intensive training in observing and managing group process. My concerns in those early years had primarily to do with herding the others, tallying to make sure all were accounted for, and ensuring that group behavior was sufficiently contained to avoid our father's ire.

My group interests shifted sharply when, at age 13, I was sent away to an all-male, Catholic boarding school where I was relentlessly bullied for the 4 years I was there, emerging with a profound fear of groups given their apparent powers to devastate. I was then keenly sensitized to being on the outside position of a group's hostility. And while still in high school, I resolved with all my being to become a psychologist with the hope of understanding the inner workings of the minds that could band together in such acts of terrorizing. Yet I became that person so absorbed with scanning outside myself for trouble that it never occurred to me to be interested in understanding myself.

As a result, when in college, I fairly quickly gave up that psychological line of work when I discovered that I could not tolerate the emotional, bodily experience of hearing what went on inside other humans. I couldn't stand that affective dimension of people's lives. It wasn't until I had a substantive psychological breakdown of my own, just before turning 30, that I finally went into my own personal psychotherapy and began to work on knowing myself. Only then could I begin listening to others.

In 1989, I decided to go back to school to get my master's degree in counseling. This is when my long-standing preoccupation with groups finally found its more formal, professional route to the study of their structural, nourishing, and destructive elements.

Up to that point in my career, I had been working primarily in the field of corporate communications and marketing, so my original intention in going back to school was to learn group facilitation for consulting purposes. But my first training placement was as a group therapist in an

INTRODUCTION

agency in Pittsburgh called Family Resources, where we offered groups for families affected by childhood sexual abuse. Walter Smith—one of the cofounders of the Western Pennsylvania Family Center where I later studied—was then the clinical director at Family Resources. He introduced us to the concepts of Murray Bowen's family systems theory, inviting us to look at the symptoms of child abuse in light of each family's broader dynamics and history. That experience influenced me to become a psychotherapist—to work with personal, psychological issues rather than with organizational ones—and much of my early work as a therapist occurred in groups. Moreover, in spite of that early exposure to Bowen, my formal group training was originally in the psychodynamic tradition, particularly in the Tavistock approach based on Wilfred Bion's ideas. I also eventually found my way to the international transactional analysis community where, in addition to being introduced to Eric Berne's writings—much of it on group and organizational processes—I also found a professional community with which I could begin discussing and developing my thinking. That led, over time, to the papers and ideas that are collected and reconsidered here.

Sidling up to the primal horde

In the past five years, I have been focused less on aspects of facilitating groups—as the designated leader—and more on my membership in them, certainly as a family member, but also as someone working in a sizably anxious institutional system, as a citizen of my possibly reactive country, and as an organism of this world. What does it mean for me to be alive on this planet, at this time? Indeed, my understanding of effective group leadership is much informed by what I have come to see as effective group membership, which I will explore in more detail in subsequent chapters.

More so, I have been interested in what groups can give me that I cannot get elsewhere, interested in what they allow me to achieve—maturationally and productively—that I cannot achieve solely on my own. Given those possible gains, how can I manage myself in the intensity of my various groups without resorting to my usual strategies of over-functioning and cutting off when the emotional climate heats up?

It's true that for all the thinking, reading, and studying I have done with regard to groups, part of my interest in them is still an emotional process—still seeking to secure an inside position, wanting to avoid expulsion, aching to run things my way. And that's where my relationship with groups is still so often fraught with hatred and fear.

Think of a time you may have seen a child frozen in terror. You may have noted an expression of intense, highly focused attention directed toward the source of that terror, as if the child was trying desperately to outwit the threat or at least to figure out how to slip as quietly and undetectably away as possible. That's how groups riveted me for many years, from adolescence on. Then, in the latter 30 years of my life, my relation to groups gradually shifted. I became less fearful, though ever-respectful of the group's greater power; less inclined to hating them all the time; more engaged by what groups could do productively; then little by little seeking them out for what I needed from them. And now, as I amble into my sixtieth decade, I begin to feel what seems strangely like love, in the sense of caring, a protectiveness and affection of the kind that likely lurked behind my anxious, harried management of my early sibling group. Yes, groups can still be scary and frustrating. But they also represent my human life, the ground of my being in this world. Whatever their disappointing shortcomings, something opens more warmly within me when I can accept my groups as they are, when I stop wasting my time wishing they were some other way.

Sometimes we arrive at this kind of acceptance and appreciation of one another as individual members of our groups. And with this can come the sense of groups as embodying the future, the ongoing nature of our more limited, individual lives.

Studying ourselves in collective life

One of the key themes of this book is that of studying ourselves—getting to know ourselves—in the human groups of which we are each inextricably a part. To "know thyself" seems to require knowing one's place and way of being as also integral to one's groups of various kinds. Even my own internal, subjective experience—that which gives me my personal sense of individual identity—is profoundly interconnected with and influenced by that larger living force that surrounds me, a force that is also paradoxically generated by me as well as by the others with whom I am grouped.

I think of this noticing and distinguishing my inner and outer systems as a form of naturalistic, observational research, which in later chapters I will link to the tradition of science as well as to the phenomenological tradition in philosophy and psychology.

For me there is an element of hope and humility in adopting this kind of research orientation. It suggests that we can face and make contact with some of the more disturbing aspects of human life in a way that we might also gradually take in and work with. And it suggests that we can let go of the idea that we will arrive at a final understanding or conception, that for every momentary sense of wholeness we may feel or articulate, the next moment will open onto a possibly wider, perhaps more chaotic vista. Our work of knowing ourselves can never be done. Yet we can take what grace might be available in our latest approximations.

This is also the spirit in which I will take up what I see as the never-ending process that is human learning.

INTRODUCTION

The human condition

I am borrowing the title of Hannah Arendt's (1958) classic in philosophy, *The Human Condition*, as a way to talk about our process of learning in relation to two key questions: (1) Where do we find ourselves and (2) What constitutes a self, our individuality?

That is, to figure out how to get to where we want to go—to follow and achieve our aspirations—we first have to know where we are: our human condition, our fate, if you will. And to know this, we begin with sensory experience, a mind–body in relation to others and to a whole world. Yet what is this body, we might wonder? What is this world in which we find ourselves? Perhaps most importantly, who are these fellow creatures among whom we find ourselves? In short, to know ourselves and our world, we must study ourselves in collective life, not as isolates.

I think there is strong evidence for the contention that humans are a highly social species (like ants, naked mole rates, and bacteria); thus we can usefully speak of human groupings (families, organizations, and communities) as natural living systems. We are a multigenerational form of life requiring metabolism, responsiveness, and reproduction; composed of ecosystemic, interrelated parts that are all in dynamic motion; both infinite, in one sense, and yet quite finite, quite mortal.

The human bodymind

Another recurrent theme in this book is the human body or, as I like to call it, the human bodymind. I use this phrase as a way to talk about the interplay within a body of its various aspects and constituents as well as the interplay among human bodies (see Chapter 2 in this volume).

For example, the individual human has a body with a triune brain whose functioning is affected by the proximity, activity, and reactivity of other individual human bodies with their minds. Thus, human neurophysiology can be described as interdependent and cross-regulating, a collaboration and conflict between physiological and neurological processes (between body and mind), and a collaboration and conflict between individual and group (between forces for togetherness and forces for individuality). All of these tensions have the potential for both productivity and destructiveness.

Even human neurological processes (i.e., perception, memory, and concept development) function collectively when we make use of others to verify what we see, to help remember what we know, and to create new ideas and tools. To give just one example, the scientific community builds on interdependent findings to construct a communal, more reliable knowledge base, a foundation from which new findings can then emerge.

As I will explain in more detail in the chapters to come, I see the lifelong, human learning task as involving the integration of the data of experience with the breakdown of that integrating process leading to dysfunction—insanity, repeated mistakes, disorientation, inflexibility, violence, war, and so on. In the words of philosopher Mark Johnson (2017), "Studying our embodiment, and all its implications for who and what we are, helps us learn how to be at home in our world" (p. 228).

Legacies: an intersection of frameworks for encountering group life

Many thinkers and teachers have influenced my approach to group work, which I will acknowledge more specifically in the pages to come. Yet I want to say here how much I am particularly indebted to the legacies of Eric Berne, Wilfred Bion, and Murray Bowen—whom I affectionately refer to as "the three Bs." Although it is outside the scope of this book to articulate more fully the areas of intersection and significant differences in their distinct frameworks, still I find it interesting that all three men served in the military during one or both of the World Wars, all came into first-hand contact with post-traumatic symptoms in soldiers and believed it important to include patients in treatment decision-making, and all three were initially influenced by and then departed substantially from the psychoanalytic theory that was orthodox in their times.

For all their differences, however, all three men, adopted a variant of the research attitude in their work and when developing their theories of which group, organizational, and family system processes were of central importance. Perhaps most importantly for me, their perspectives attracted individuals who formed learning communities around those ideas and practices with the hope of extending those initial conceptions. Within the transactional analysis community, for example, Oded Manor (1992) offered his own perspective on these three traditions and their interconnections in his article "Transactional Analysis, Object Relations, and the Systems Approach: Finding the Counterparts."

These communities have, in turn, been actively present to nurture my own growth and thinking about human life in groups.

An intuitively developing vision

The chapters that follow appear in the sequence in which they were written over the course of nearly 15 years, during which all were either published as articles or given as presentations. And although they were not written with any conscious, overarching vision or narrative in mind, I am keeping their chronological sequence for this book because the chapters tell the story of a gradually developing perspective on group life, a story of the kind that I believe many of us could tell based on our own experiences, even though the details and manner of expression would certainly vary significantly.

INTRODUCTION

As I look back over the unfolding of these particular papers and presentations, it is as if I were writing a story I did not know I was writing. I can now see connections among these papers that I did not realize were there at the time of their composition, much as we each may come to the archive of our collected and collective materials with a view and understanding that is renewed with each new reading. Today I can see the discernible pattern that I believe progressively helped me in my group encounters and gave me the framework I find so useful today. And what, I wonder, will I see when I read this tomorrow?

It is more obvious to me now, in retrospect, that as I wrote, I was integrating the various theories I was being taught while also developing my own perspective. This more personal view came as a result of steady involvement with groups of many kinds—therapeutic, training, work teams, and families. And my sense of this integrating led in time to my interest in learning processes, for both individuals and collectives, giving me a sense of the interplay and influencing that occurs between the neurophysiology of individual humans and that of human groups, our social species, on our way to learning our world.

Today, for example, when I find myself overwhelmed with the data of a particular group's life, I can more easily and intentionally remind myself to step back, to adopt an attitude of what I call *engaged researching*, thereby giving myself room not to know but to trust in this basic human capacity of reception. It is a place to begin that is always reliably there.

I am then more able to look more closely at the emerging details and nuances of what I have called the *shared bodymind*, that is, the conception and awareness of the intimate, complex interaction between the body and mind—the neurophysiology—of the individual person and the dynamic interplay and mutual influencing among multiple individuals comprising that organism called a human group.

My next endeavor led me to look more closely at my experience of *learning and hating in groups*. With a now greater capacity for tolerating the intensities of group life—part of which required a conceptual scaffolding—I could look more honestly at the degree of my hatred of groups and my concomitant destructive strategies for discharging that hatred, which, of course, proved to be shared by many of my fellow group associates. At the same time, I could begin to see and feel for the first time what groups could offer by way of extraordinary benefit, especially in the form of the learning that has been so important, albeit challenging, for my life.

Over time, my understanding of the neurophysiological interplay in groups has been informed by emerging research in neuroscience. In particular, I became intrigued by what I saw as a recurrent dynamic in human relations, that of the *social pain response* identified by a group of social-cognitive researchers in the early part of the twenty-first century. This idea offered me a way to conceptualize a common social system process that can appear, on the surface, highly variable, often showing up in explosive forms. Yet beneath these outer manifestations, I began to see a recurring underlying pattern, and that offered me a means of languaging such situations when they emerged so heatedly in groups and in my life generally.

With this increased bodily and conceptual capacity, I turned to what I called the process of *looking for trouble in groups*, especially those that were explicitly intended for professional development and growth. I could now deliberately seek out the areas of greatest difficulty in my groups—making active contact with those areas of trouble—which allowed me and the group to engage with what I came to see as the most important areas for growth and learning for a particular group and its members.

Over time, I began to conceptualize this generative learning process in terms of *maturing as a community effort*, an interplay not just of our disorganized neurophysiological states but also of our movement forward in integration, meaning-making, and increased bodily and mental capacities for encountering life.

All of these stages of my development have now brought me to thinking in terms of *groups that learn (and groups that don't)*. This has greatly expanded my conception of human learning processes, which, in turn, has informed the principles and practices that can potentially facilitate such group learning, such collective and individual maturation, to borrow from Winnicott.

Can we each be curious about the sequencing of our own histories—as individuals, families, organizations, communities, and other abiding human collectives—as a means of discovering the patterns at work outside our awareness and sometimes, by providence, simply the result of natural, life-supporting developments in time?

That is the meaning that emerges for me in the rereading of my own work. And you, in reading as you will, may make other patterns and meanings, may see things I missed, and will surely bring a different sensibility and legacy to bear. That, to me, is one example of what can make community life so generative. The clearer I can become about my own perspective and the more I can put it into some form of expression, the more I can then welcome, even eagerly solicit, the varying views of my colleagues, friends, fellow citizens, and family members. I am grateful for being heard and, at times, corrected.

One thing I like about this book being a collection—written over time, with no overarching plan—is that it will necessarily be incomplete and will spare me the compulsion to write a single, fully integrated theory, which in any case I could never do.

I also have a sense of needing to approach this as I would the making of an artwork, with less expectation of consciousness and more alert to the signs of an emotional, aesthetic impact, more respectful of my own more spontaneous expressiveness, the current limit of my maturing.

1

ENGAGED RESEARCH

For many years, my professional involvement with transactional analysis was limited to my affiliation with TA-trained colleagues. My own work as a psychotherapist and human relations consultant was informed almost exclusively by psychoanalytic theory, although in later years I increasingly studied and used Bowen's family systems theory in my work with individuals and organizations. As I struggled to integrate these different strands of my professional life and training, my interest in and contact with the transactional analysis community grew substantially. I was especially drawn to the community's efforts to foster a practice that cut across traditional disciplinary boundaries.

One colleague, for example, had expertise in transactional analysis, psychodynamic psychotherapy, and group process. When she mentioned that she would be conducting a three-day workshop for advanced, not-yet-certified transactional analysis trainees, I asked if I could observe. I wanted to understand how her attention to and interpretation of group dynamics could support individuals in learning about the problems encountered in their clinical and educational work. She agreed to my request, as did the training program's director. My research focus was to be the effects of the trainer's practice on the group.

Twenty-four trainees were present. As a group, they were culturally and professionally diverse—counselors, psychotherapists, educators, and organizational consultants. Their depth of professional experience varied considerably. They were asked to present points of impasse or exceptional difficulty in their ongoing work with a client or organization. This was to be examined in light of parallel process, transference, countertransference, and developmental and character issues. Demonstration supervision would address experientially the difficulties group members were having on the job.

The trainees were not given prior notice of my presence as a researcher-observer. The first day, I simply introduced myself and my background and described my interest in being there. My contract with the group was to remain silent, whether taking notes or sitting there. Group members were shyly welcoming. One expressed appreciation for the fact that someone had an interest in researching transactional analysis.

I saw some individuals on occasion watching me, although infrequently and eventually not at all. A few kidded me about taking notes. A few asked questions, during breaks, about what I was noticing or thinking. But the workshop schedule was so tight that there was not time to answer other than in broad generalities. Between group sessions and at the end of each day, I talked at length with my colleague, the trainer, about what I was seeing and what she was noting. That was the basic setup.

What's going on here?

I have, over the years, conducted many intensive groups. Compared to those experiences, this workshop proved unremarkable in its dynamics. So, I was unprepared for its intense emotional impact on me. Freed of the responsibility for facilitating the group's process, I wondered, why would my affective levels be so much higher than usual?

My focus shifted immediately away from the trainer's behavior to the palpable difficulties being lived out, in the group, by the participants as they presented their challenges with clients. I became so immersed in the process that I often had no idea what I was thinking. For the first day and a half, I wrote almost continuously, noting what caught my attention or generated questions. I filled pages with these fragments and free associations. Then, for the second day and a half, I sat without writing, as if I had already captured all I needed, yet all the while finding myself more and more exhausted. It was like endurance training, far more demanding than having an active role.

On two occasions I did not restrain myself and made statements about what I saw in the group. I could tell by the puzzled faces of the group members that they did not understand what I was saying and were confused about my speaking at all. Although my contract violations seemed to create no lasting problems, I wondered how my "outbursts" might have been experienced in a less sturdy group.

I could not tell what I was learning. I could not remember why I was there. Keenly engrossed, as I was, with how the culture of transactional analysis hindered and helped its practitioners and how the practice of helping actually worked in the untidy midst of human relations, I still found the 3 days to be endless and felt I would burst with fullness. I was both committed to being there and itching to run from the room.

I gradually became invisible to the group, in part because they had so little interaction with me, in part because I presented no apparent threat, and largely because of their increasing engagement with the material and learning in the group.

Yet during the participants' closing remarks and feedback, I was startled when one of them addressed me directly by saying, "I thought at first you were going to watch us, like we were bugs under a microscope. But you were very respectful. I didn't feel like you were watching us as much as watching out for us." Many heads began nodding in agreement accompanied with warm smiles.

Whether from exhaustion or from gratitude for this man's interpretation —for his effort to give meaning to something I had not seen—tears came to my eyes. His comment helped me feel what I had been living outside of my awareness. But still I had no idea what had actually happened. I had only the sense of discovering something important, something I had not consciously set out to find.

The risk of gathering data

One of the paradoxes of professional relationships with clients, students, or others who come to us for help is that we are supposed to maintain our objectivity even as we, too, may become infected by what the person has brought us to work with. This is an inevitable consequence of our agreement to listen to and think with someone emotionally troubled. We literally open ourselves to the impact of their affective world. Something of their experience comes to reside uncannily within our bodies and selves.

Psychoanalyst Wilfred R. Bion (1897–1979), throughout his life and writings, acknowledged the problem inherent in this way of working:

> In turning ourselves into receptors we are taking a big risk. From what we know of the universe we live in some of the information may be most unwelcome; the sound or signal we receive may not be of the kind that we want to interpret, to diagnose, to try to pierce through.
>
> (Bion, 1980, p. 60)

Yet this risky procedure—of receiving what we may not want to know— is precisely how we come to be of help. We do not just apply our cognitive capabilities to sets of words or facts; we figure out how one can live within the same emotional forces that are disrupting another's life or capacity to function well. It was Bion's insight that the psychoanalyst or other helping professional first processes the client's unconscious material at a bodily, felt level, and then essentially works up from the body into the mind, where that experience can be symbolized even as it remains bodily rooted.

Bion's work and contributions to psychoanalytic theory were familiar to Eric Berne, in keeping with the latter's awareness of and interest in other psychoanalytic writings of his time. In *The Structure and Dynamics of Organizations and Groups* (Berne, 1963), for example, he noted:

Bion is one of the few people who have tried to observe what goes on in a group from a naturalistic point of view, not trying to prove or disprove anything but merely asking themselves: "What's going on here?" In some ways Bion's work is more interesting than the usual commonplace statistical studies.

(p. 102)

Berne's link between Bion's thought and attitude of research into human phenomena helped me understand that one of Bion's key concepts—that of the containing function—might describe more than just the responsible and useful behavior of a parent, teacher, mentor, consultant, or psychotherapist. It might also elucidate a process whereby research need not interfere with treatment nor just evaluate its outcomes. Conceived in light of containing, I wondered if research might itself perform a therapeutic or educative function that would enhance our work as professionals in a variety of fields.

Bion's containing function

Among the many contributions made by Bion to psychoanalytic theory and practice, perhaps best known is his concept of the *containing function*, which he developed over the course of his career in such works as *Learning from Experience* (1962/1977c), *Elements of Psycho-Analysis* (1963/1977b), and *Attention and Interpretation* (1970/1977a), to name a few.

As Bion used the term in his many writings, "containing" has a special meaning. It denotes a series of critical steps in an interpersonal process. He proposed thinking of this process prototypically in terms of the mother–child dyad, a proposal that drew initially on the work of Melanie Klein. He theorized that the child's unorganized, unconscious bodily experiences of being alive in the world would be communicated nonverbally and concretely to the mother. The mother would receive such communications primarily at the body level. In ideal circumstances, she would sit with the unsettledness or discomfort picked up from her child, generally corresponding to the child's highly charged experience of hunger, fear, fury, or excitement. The mother would then respond, after reflection, in a manner that would relieve the stress (for example, by providing food or attenuating the alarming stimulus) or that would name it (perhaps by saying, "I know. You don't want me to put you down right now. But I have to go do something and will be right back"). Through her act of reception, reflection, and response, mother conveys a felt sense of being able to bear lived experiences rather than needing to push them out of sight and consciousness, like so many unpleasant thoughts.

To illustrate, I remember when my then-infant daughter was first being weaned from her mother's breast in readiness for her mother's return to work, sooner than any of us would have preferred. It was my job to get our baby used to the bottle. Her initial response was utter fury and noncompliance. And based on that display, I had the distinct impression that I was murdering her, crazy as that may sound in retrospect. I was so caught up in the moment and in the passion of her protest that I believed the worst. I recall the nearly overpowering urgency I felt to hand her right back to her mother, calling off the whole awful mistake. But then, looking down at my livid baby, it occurred to me that she might be feeling as if I were killing her by introducing this rubber-tipped, plastic object filled with formula. Of course, I knew it was not my intent. I also knew the transition would not kill her, although I felt deeply sympathetic to her distress, loss, and anger. With this awareness, I felt a distinct shift in my body, in which I held side by side both sympathy and the confidence that she could adjust to this traumatic change. That is, I was able to contain my initial bodily panic and fantasy (of harming her), reflect on the possible source of those feelings, come to some sense of accord within myself, and then convey that balance to my daughter. No, she did not immediately calm down. But I was able to remain present in a way that eventually allowed her to adjust, in her own time, unimpeded by my own reciprocal hysteria. What I remember most was the feeling of having awakened from a dream or mini-psychosis, accompanied by a sense of sturdiness at having separated myself while remaining in contact with her—engaged but not entangled.

Bion suggested that in a normal process of development, this sequence of communication, reception, reflection, and considered response helps the child learn to process her or his own experiences. Furthermore, although Bion specifically referenced the mother–child dyad, he saw this as a universal interpersonal exchange between any two or more intimates. In particular, he was concerned with this communicative interaction between the psychoanalyst and patient, which exchange bore directly on the success of treatment. As with the mother in relation to her child, the analyst or other helping professional could demonstrate that the client's internal emotional world could be suffered realistically rather than avoided and elaborated into catastrophic or magical "phantasies."

The term *phantasy* is used by Kleinian analysts to indicate an unconscious narration of bodily experience that differs from our more conscious fantasies or stories. An infant, for example, may link the experiences of hunger-then-gratification (or separation-then-anguish) with the clustered feel-smell-sound of her or his mother. Based on the infant's own temperament, as well as the rhythm developed with the mother, a story or hypothesis about certain feeling states (hunger, gratification, contact loss, pain, Mommy) will be developed in which the infant's sense of autonomy and helplessness will play a role in connection with that of the powerful (giving, withholding, loving, or hateful) parental figure. Even more, the infant will begin unconsciously to live this story about what it means to be alive. These unaware and archaic qualities of phantasy are similar to those of script and protocol in transactional analysis.

However, phantasy also differs in that the narrative arrangement is not impressed on each of us but actively created, outside awareness, as a function of our human predilection for making sense of sensory experiences. Furthermore, we make these unconscious bodily stories not only as infants and children but throughout life. And as with script and protocol, phantasies operate as more than just simple explanations we create but also as templates that guide our behavior and choices.

Receiving

In his conception of containing, Bion meant more than just biting one's tongue. He described containing as a sequence of receiving, mentally processing, and speaking or interpreting. It is not easy work, and each step in the process is equally critical. A therapist, consultant, or teacher must first be receptive to the communication from the client or student, which inevitably comes loaded with emotionality. That emotionality or affect must then be held and mentally chewed over—lived within and digested—in order to understand the nature of the emotional communication. At that point, it becomes necessary to speak to the truth or fact of that communication, no matter how hard it is to hear and know.

We often think of communication as occurring through words or through developed nonverbal symbols (images, gestures, sounds, or touch). Bion and many other psychoanalysts, however, have been concerned with communications that are sent and received unconsciously, outside ordinary channels of awareness. These communiqués are typically states of bodily affect, the ways we register life's impacts as well as the surges within our own dynamic bodies, similar to Daniel Stern's (1985) concept of *vitality affects*. Some of these affects can be subtle; others arrive with the full force of a storm. And yet they can remain undetected by our minds, especially if the emotions correspond to experiences for which we have no ready words or shared symbols. Entire swaths of life can exist in a limbo of unexamined experience. Bion termed these *beta elements*, the bits and chunks from which we eventually hope to structure our consciousness. As psychoanalyst Michael Eigen (1998) put it, "[Bion] calls on us to face the fact that our ability to process experience is not up to the experience we must process. This is not only so in infancy, but all life long" (p. 99).

Human beings are inclined to put unpleasant experiences out of mind; in fact, such experiences may never reach consciousness. They are repressed, split off, relegated to the outer limits of the bodymind. Yet, as with offshore dumping, these repressed experiential contents have a nagging way of floating back, muddying the waters of our status quo. Sometimes, too, these rejected experiences find their way into the bodies of those close to us, as reverberations transmitted through a process that psychoanalysts and others have termed *projection*. That term captures the force with which these orphaned bodily experiences can be aimed to hit their targets, but it also has the unfortunate effect of suggesting that we are broadcasting to one another along invisible radio waves. It is more accurate and simpler to say that, as human beings, we are acutely sensitive to one another: to tones of voice, to minuscule facial twitches, to muscular tensions, to smells, to changes in breathing. Countless cues operate outside awareness to signal changing states in a person, changes that can be picked up nearly instantaneously by someone in proximity and even relayed in a flash throughout a group, community, organization, or family. Affect is extremely contagious, especially when burning.

I worked once with a man, a physician by profession, with whom I used to find myself moved to tears in the middle of his sarcastic and complaint-filled stories. At those moments when what I felt intensely did not match his words, he typically responded to my inquiries by claiming to feel nothing, a claim his body language seemed to support. He looked at me as if I were crazy. Yet over the course of several years, we were able to link my feeling states to his aggressively repressed inner life. Gradually, he began to cry. I was then free to feel with him, not just for him. Eventually, he found these words to describe his everyday work with his own medical patients:

> If I get overwhelmed, if I have a patient I don't know how to help, nothing is in my brain. They want help. I draw a blank. It feels bad. It begins to feel out of my hands, like a patient I can't understand. If I can't make the connections, I don't want to deal with it. I make it bigger instead of figuring out how to connect. This "out of hand" thing is like knives, like something painful. I take that part of my body and put it in a box. I don't want the pain but still have to go on as a person. When things get out of hand with my patients, I start thinking it's about me. I'm stabbing myself.

He not only portrays an inner world greatly at odds with his outer, apparently confident professionalism but also details his struggle with containing the anxiety and upset picked up from his patients. Like all of us who are similarly affected, his strategy is to draw a blank and place problematic parts of his body into figurative boxes. His words illustrate how our internal experiences become converted into literalized objects, like knives, and how we lose our sense of who is stabbing whom.

Processing

A person's ability to suffer both pleasure and pain, and thus to render those experiences meaningful in the larger context of life, performs what Bion called the *alpha function*. This human capability involves creating mental and bodily structures that allow the shocks of life's beta elements to find a home within a larger system of meaning. Religion, for example and for all its many imperfections, stands as a brilliant human creation, a product of alpha function that attempts to order and make sensible life's sufferings, from ordinary misery all the way to disabling loss and ecstasy. As Eigen (1998) wrote, "Our religions and psychotherapies offer frames of reference for processing unbearable agonies, and perhaps, also, unbearable joys. At times, art or literature brings the agony-ecstasy of life together in a pinnacle of momentary triumph" (p. 101).

Alpha function, therefore, is more than just finding a set of words. Unless those words offer us a living symbol, they can be used to mean nothing, to discharge affect, and to obscure what may actually be happening. As speaking animals, we can be quick to vocalize without really thinking and to use words to toss back or reproject the other person's disturbance. Speaking then becomes an automatic habit that is particularly problematic when we have agreed to work with someone who is experiencing difficulty. Yet under the guise of professionalism, we may expel what we are picking up unconsciously before we are even aware that we are doing so. And we may do so simply because we have not learned to endure and reflect on the more passionate and disturbed states of being human. As Bion (1980) tersely noted:

> It is too often forgotten that the seriously disturbed patient is being disturbed because he is aware of something serious, even if his analyst isn't, and does not want to be reminded of it. Analyst and analysand can be at one in wishing to deprecate the seriousness of "mental pain"—hence a dangerous collusion.
>
> (pp. 117–118)

As part of this dangerous collusion, we may be particularly prone to eliminating what makes us feel crazy or psychotic inside. The unfortunate result is that we then cannot begin to demonstrate the kind of understanding or thoughtfulness our clients or students need to learn, through reintrojection, in order to live their lives most fully. Bion cautioned against prematurely acting on or reacting to an emotional communication received with great discomfort or anxiety on the part of the professional. Rather, speaking was to take place in a relative state of neutrality, which signaled the fact that the distressing emotions could be lived within instead of continually tossed to the far side of one's psyche.

Interpreting

Once alpha processing has taken place, Bion's third step of containing involves delivering the interpretation. The interpretation is the product of that processing, made with one's whole mind and body from the emotionality that has been communicated, outside awareness, by the other person.

To interpret is to introduce that processing into the interpersonal realm, to reference the intimacy of the exchange.

Interpreting can be written about more easily than actually conveyed. The moment of conveying or speaking may be fraught with a new set of emotional tumults, typically related to the ways we have historically interacted with others. To speak what has been repressed is to open the subject—both topic and person—to conscious distress. It can be upsetting finally to know and say what has been going wrong, without the luxury of pushing it from view.

Speaking what one experiences also signals the presence of a separate mind that, in silence and reverie, can otherwise seem to be one with one's own. This evidence of separateness crushes fantasies of psychological and emotional merger. These and other perils of making a meaningful interpretation are described beautifully by Robert Caper (1999) in his book *A Mind of One's Own*.

Because interpretations are often verbalized, we can easily become lost in their content and in questions of their accuracy. These issues are relevant. Yet an interpretation must first be validated on the basis of its function in an interpersonal process. Being accurate is of little use if the spoken words are meant mainly to discharge affect or to interrupt an unnerving containment and processing, as when I interjected my non sequiturs into the training group. The intent of a vocalization will be felt well before its literal meaning. In fact, we can think of meaning as being built first from relational interactions—what it means to be in one's body in a world of others—before the content can elaborate that experience of being and relating.

Remaining silent might therefore seem the safer course. But for Bion, it is not enough just to receive and sit with an experience. The interpersonal cycle is not complete until the results of that reflection have been brought consciously and courageously into the realm of human exchange. They have to be lived and relayed, in relationship, to be of transformative use.

Living with the facts

Let us say that the desired outcome of a therapeutic, educative, or other developmentally enhancing process is an increased capacity for being oriented to reality and to the facts of one's life, with a resulting improvement in social and personal functioning or adaptation. This definition should hold true for cognitive-behavioral therapy, teaching, family systems coaching, organizational consulting, and psychoanalysis. It would certainly fit the transactional analysis claim that the purpose of therapy is to support spontaneity, awareness, and intimacy.

Our methods for bringing about such improvements differ among various professions and practices. But if reality orientation is a function of bearing the degree of emotional intensity and conflict associated with our human condition, then regardless of treatment or intervention method, what facilitates that success is what Bion called containing—not just by the professional but ultimately by the person seeking help. For according to Caper (1999), "Proper containment should not only help a patient bear a current state of mind, but also help him to better bear future ones without help from an external object [or significant other]" (p. 141).

Just as a historian must construct a narrative out of pleasing, ugly, dull, and even disjointed data, we each attempt to bring together our own relations to what Bion saw as loving and hating, self and other, knowing and not knowing, and other such unbearable tensions. But even with the best of help, we do not all recover. My daughter might have decided never to be weaned; my physician client might have decided to keep his internal knife-work hidden and deadly; I might, in frustration, have left in pieces the analysis of my experience observing the training group.

Our power as professionals is inherently limited. Berne (1966) referenced this humbling fact by quoting a famous physician's retort when praised for a patient's remarkable recovery: "I dressed his wounds; God cured him" (p. 63). Berne stated emphatically: "The therapist does not cure anyone, he only treats him to the best of his ability, being careful not to injure and waiting for nature to take its healing course" (p. 63). In a similar spirit, Caper (1999) pointedly stated, "The fate of the analysis is determined ultimately not by the analyst's interventions, per se, but by the dynamics of the patient's unconscious" (p. 19).

Facilitation, in other words, is not the same as ownership. I may enter another's psychological world, intentionally or unintentionally, but what that person then does with the result of my understanding is outside my control. So whether we want to call improvement the result of God's grace, nature's healing course, or the patient's unconscious capacity to embrace life, Caper and Berne are advocating humility and reserve as accompaniments to discipline in professional practice.

Applying a bandage as expertly as possible or delivering a fully contained interpretation as neutrally as possible does not necessitate coldness or distance. Rather, as professionals, we arguably give our best when all of our resources are focused on what helps most. For Bion, optimal treatment of cognitive-emotional disturbance took the form of his living, in the presence of his patients, a steadying relationship to what could be described as his patients' inner lives. Such balance is consistent with a researcher's efforts to study a situation and to deliver dispassionately her or his most complete results.

Engaged researching

As just mentioned, calmness and reserve need not be equated with emotional distance, cutoff, or objectification of the other person. The latter are defensive, antitherapeutic stances. They signal the professional's refusal to contain the cognitive-emotional data that can facilitate a solution to the client's difficulty. Only by making emotional contact with the experience of the person who has come for help can a professional receive, sustain, and organize complex, sensory and emotional data.

Furthermore, research that demonstrates such engagement and containing does not objectify the subject, in the sense of turning the person, group, or issue into an object or thing to which one has no ethical relationship. Rather, engaged research acknowledges its affiliation with or participation in what is being examined. Objectivity, therefore, cannot be characterized by distance or emotional cutoff. We can only see the facts of our human condition clearly when we acknowledge our own relationship to them and achieve enough neutrality to face and understand what we are living within.

Robert Sardello (1971), in a paper on research in phenomenological psychology, wrote, "The ... most primordial meaning of objectivity ... is an attitude of respectful openness to the whole of our existence, which allows, through our involvement in the world, reality to reveal itself the way it is" (p. 64). This faith in the importance and usefulness of seeing the facts of the world realistically was certainly endorsed by Berne. Similarly, Caper (1999) suggested:

> Analytic containment is not an effort to make the patient feel relieved, or good about himself, but only to help him think and feel what is true. The relief comes from the fact that the truth is always less persecutory than the phantasy that had displaced it in the patient's mind.
>
> (p. 42)

This potential for reality's relief is also portrayed by psychotherapist George Downing (2008) in conjunction with his use of videotaping to help troubled single mothers see how they interact with their children. The video footage is a starting place for thinking about what is going on in those interactions and what might be done differently. Videotaping is also a common research tool. It can, like photography, lend itself to objectifying or shaming what it captures. But Downing, in his writing, presents his clients in a manner that makes it possible for us to identify with their dilemmas as parents to young and sometimes difficult children. When examining the video images with his clients, he balances their problematic behavior against the very real challenges of raising a child. He contains the situation whole. In discussing what he sees with his client, he models a collaborative, inquiring approach that the mother can then learn and live with respect to her baby. Downing, in effect, uses a research tool to support his clients' capacities for objectivity and contact with the emotional difficulties of parenthood. From there they can begin to make different and often more satisfying behavioral choices.

For research to have an effect that is growth enhancing, the researcher must be engaged and participative in the process of understanding the emotional phenomenon being studied. It requires the use of the researcher's full being.

Reflections and further questions

In the year following the training workshop, I learned that my role as researcher-observer generated several tangible outcomes. The trainer, in addition to receiving support during the workshop itself, reported understanding some of her key didactic material differently, which allowed her to modify it. For example, I had noted the way the concept of *parallel process* was invoked often in the group and applied in ways that continually allowed the term's meaning to slip. Identifying a "parallel process" frequently seemed to be a pleasing cognitive activity for the group, one that involved looking for and matching aspects of the demonstration supervision process with what the supervisee had described about the experience with her or his client. There seemed to be no instance in which a match could not be found. But use of this "matching game" then overrode other possibilities. It shut down thinking and further exploration of difficult client interactions. Use of the concept appeared to meet the group's need for interrupting emotional involvement, which I could well understand given the affective intensity in the group. At the same time, I hypothesized that the term also was needed by the group—perhaps by the larger transactional analysis community—to describe certain clinical phenomena that were not otherwise sufficiently accounted for by the group's predominant theory. In that sense, use of "parallel process" seemed to indicate the group's ambivalence about affective interactions between client and professional (or between supervisee and supervisor) and about the "messiness" of psychic or mental processes. It was as if to say, "We are naming something we see, but we don't know how to get much closer to it." I wondered to what extent this might have been inherited from Berne.

I also noted similar confusion around terms such as *transference*, *unconscious*, and *interpretation*, including what seemed to be the unnecessarily obfuscating way these terms are often used or defined within the psychoanalytic literature. Observations of these kinds allowed the trainer to clarify definitions and to specify the ordinary human phenomena to which they refer. She was also able to bridge more carefully between the cultures of transactional analysis and psychoanalysis. That modified material, she later reported, was well received by subsequent trainees. She and I were also able to collaborate on an unpublished paper that gave feedback to the group of trainees I observed and that also developed some of the workshop ideas that had not been fully explored. An abbreviated and edited version of my process observation notes was shared with the training program's director and faculty, which generated a lively exchange over a period of several months, possibly to affect future curriculum planning at their center.

Yet I entered this particular group with no intention of playing an active role. I was just curious. I imagined my curiosity to be of little consequence to anyone else. But by engaging as I did with the group, I sat both with my notes and with a level of emotional information that I could barely tolerate. And as Bion might have predicted, my doing this in and for the group contributed to a climate of thinking about difficult emotional work with clients. However, contrary to what Bion might have suggested, I did not deliver an interpretation. Or rather, the interpretation—the words that described an unconscious process—came from someone else in the group, not from me as the researcher/analyst. Perhaps in modeling what was being taught about working with difficult states of affect, I demonstrated those very principles. I wordlessly conveyed the therapeutic effect of such containing in our work with clients of various kinds. Might that containing, I wondered, be the therapeutic effect of engaged researching?

Reflecting on this group experience has shown that this kind of research has been an integral yet unconscious part of my work life for years. Perhaps it is so for many of us in the human relations professions. After all, as practitioners, we are asked to define the desired outcomes of our work. We hypothesize how to bring about those outcomes. We strive to practice conscientiously, aiming for clear contracts and attending to the effects of our interventions. We also willingly modify what we do to bring it more in line with contractual objectives and intervention efficacy. In that sense, we already operate with a research model that tests the effects of our work against our expectations, in keeping with Berne's emphasis on working scientifically.

Such evaluating is also very much in the spirit of action research, as conceived by Kurt Lewin (1948) and developed by many others, particularly in the field of education. Like action researchers, we do more than simply test our effectiveness. We modify how we work in accordance with the continuous feedback we receive in the form of our research data. We engage in an iterative process of action, evaluation, and modification that ideally fine-tunes our practice as it achieves our work. We learn from experience.

Nevertheless, I find it difficult to live this engaged way of working that appears to pervade my professional life. For when gathering data on any human system, I begin each project in a state of cluelessness. I am affected by an initial chaos of emotional and dynamic information that is often unpleasant. When I unthinkingly evacuate that intensity—as if reflexively gagging—I intentionally have to take the affect back in, to bring myself back into contact. That is typically the nature of my contract when I have agreed to understand and work with an individual or group.

This agreement can be especially difficult to sustain with large, complex, and intensely emotional systems. My work with government clients, for example, often leaves me feeling paralyzed and sick. Still, I am aware that on the other side of such illness there can be a capacity for thinking and for making decisions that function more effectively for me and my community. I find a sense of hope in knowing that I can think determinedly within and for my community, as my teacher, Elizabeth Minnich (personal communication, 15 January 2001), described her teacher, the philosopher Hannah Arendt.

Our capacities for containing—for residing within and serving our communities, families, and groups—can be developed gradually. This emphasis on learning through researching was also advocated by Gianpiero Petriglieri, then Vice President of Research and Innovation for the International Transactional Analysis Association (ITAA). In his open letter to the association's membership (Petriglieri, 2004), he also urged us to locate the areas of mess or cluelessness in our lives, to set off exploring them, and to report back to the group what we have found. His question "What do you want to learn tomorrow?" aims right at the heart of research and at the most realistic hope we can have of helping one another.

Bion hypothesized that, in addition to instincts for loving and hating, we humans are equally driven in our lives by a desire to know ourselves and others. Research into human systems, when engaged and motivated by learning's desire, brings about its therapeutic or beneficial effects precisely because it enters into those systems. It specifically affects their functioning by injecting thoughtfulness and curiosity and by struggling for objectivity. These are key interventions. The more thoughtful and objective we can be about our situations, the better and more ethically we can take action, whether as professionals or as clients.

Engagement through an attitude of research is thus most profoundly a contribution to thinking for and within one's community.

2

THE SHARED BODYMIND

The troubles that bring people for help and that make helping difficult can usefully be traced to two basic, interrelated human challenges: (1) our capacity for thinking what is not necessarily real and (2) our incapacity for bodily affect and responsiveness to living.

The first problem is that we are capable, as humans with minds, of thinking things that are not in touch with what we would individually or collectively consider reality. Yet based on such false certainties, our minds make decisions, sometimes "without conscious thought on our part" (Berne, 1947, p. 62). Often we cannot see that those decisions do not square with the facts of our lives. This shows most dramatically in individuals diagnosed as psychotic. But even at a community level, in my hometown of Pittsburgh, for example, I witnessed distorted thinking for years when ordinary people dealt with the trauma of the steel industry collapse by talking as if the mills would reopen any day, even when those very mills were being demolished. In far less dramatic ways, wishful thinking, denial, and paranoia can be part of everyday life. Berne's central focus in transactional analysis was to wake us up to the delusional aspects of our decision-making in order to place ourselves more securely in the world we actually inhabit.

The second interrelated problem is that we simultaneously live within physical, responsive bodies that can barely process or tolerate the stimuli of life. Certain facts can shock us. Certain encounters with others can leave us nearly faint with anger, shame, or fear. Most of the time we develop some degree of familiarity with or tolerance for the way we experience the world. Our daily, second-by-second responsiveness usually falls into the background, out of awareness. But in some situations we can feel with such intensity that it is as if we had a wild animal within us, "a Panther in the Glove" of our own skin, as poet Emily Dickinson (1861/1976, p. 112) depicted it. That intensity of affective response can dominate our decision-making without our being aware of it. Yet the roaring in our bodies can rarely be thought away, even if we sometimes believe we can do so.

In emphasizing that these mental and physiological dilemmas are shared by clients and professionals alike, I want to underscore the commonalty of our challenge. I want to support the already strong tradition within transactional analysis, psychodynamic practice, and family systems theory that asks us, as professionals, to resolve our own psychological difficulties for both ethical and practical reasons. I also want to contribute to the growing awareness of the relational aspects of professional work (Cornell & Hargaden, 2005, 2020). In this view, human mental and physiological challenges are reciprocally conveyed and literally shared as we work with the clients, students, and organizations coming to us for help. Self-analysis is, therefore, more than just the practice of correcting errors after the fact or of doing personal work to prevent them in the future. Self-analysis is an essential technique that guides our moment-by-moment interventions. The more we know of our common human vulnerabilities, the better conditioned we can be for the often-turbulent encounter of helping or facilitating.

I find it useful to conceptualize my work—as a psychotherapist, teacher, and organizational consultant—in terms of mental and physiological challenges. But my actual experience is rarely that neatly ordered. In fact, most of it happens outside my awareness.

The unitary bodymind

For diagnostic and intervention purposes, I often think of mental and physiological processes as distinctive. I trace their different manifestations; I decide which aspect to highlight with specific clients or students. But I am also aware that these aspects are not separate. They are dynamically interrelated—twin features of the same unitary bodymind. As Bion (1976–1979/2005) wrote, "The individual has to live in his own body, and his body has to put up with having a mind living in it" (p. 10).

To reference this more multifaceted human being, I use the term *bodymind*, as do practitioners in the fields of bodywork, mind–body integration, and holistic healing. More importantly, I choose the term to describe a form of embodiment that, in every moment, is alive to the world and making some sense of it. Like every other form of life, we are physical, sensate, and cognitive to the core—"nervous tissue" in Berne's (1964, p. 56) language. But we are alive to the world in particular and peculiar ways given our human form. We can think in highly useful and sophisticated ways; we can also deceive ourselves. We have adapted to life in numerous environments and conditions, and yet sometimes we can barely absorb the emotional data of living.

The bodymind tension in human relations work is not the same as the classic mind–body problem of Western philosophy, in which the mind is accorded a sphere of its own in which to reconcile itself with a separate physical realm. John Dewey (1925/1981)—philosopher, psychologist, and educational reformer—attempted to redress this problematic dualism by coining and elaborating the term "body-mind" (p. 217). Likewise, the growing consensus of the scientific, therapeutic, and consultative communities is that our minds are not separate from our bodies, that our cognitive capacities are integral to our physicality and vice versa. In Johnson's (2017) words, "Mind is an evolutionary accomplishment that cannot exist without a body in continual interaction with its world" (p. 40).

Even Freud, over 100 years ago, operated with this kind of psychophysiological conception, as described by Fast (2006):

> In [Freud's] ... more radical view ... the bodily in the mind refers to the mental representations of actual (bodily) experiences (e.g., in family life). Their primitive expression in the primary processes is gradually replaced by mature thought in which such patterns, variously articulated and modified, constitute the mind. The bodily (the subjective and emotional), therefore, is not restricted to primitive thought, but pervades the mind at all levels of development. The mature mind, not a body-free structure of logic and rationality, is a *body-based organization of increasingly sophisticated and nuanced patterns of experience* in which the personal and emotional are of continuing importance. The logical and rational are of major importance but are, nevertheless, secondary.
>
> (p. 275, italics added)

Phenomenologically, we may experience ourselves as split into a mind and body that do not always seem related. Mentally, we can inhabit states that appear wholly disconnected from the physical world; we can lose ourselves in daydreams, fantasies, memories, plans, fears, and dissociative states. We can also relate to our physicality and responsiveness to the world as if overtaken by forces that seem alien to us, for example: inexplicable dizziness after an interpersonal encounter, sobbing seemingly unrelated to any conscious thoughts or actual experiences, elevated heart rate and breathing difficulty associated with certain ordinary stimuli, or paralyzing dread of apparently simple interactions with others.

It can be difficult to see these disparate mental and physiological phenomena as aspects of the same, unitary bodymind with its interplay among sensing, reacting, patterning, and intending. In fact, when we speak in terms of higher mathematics or poetic creation, it is easy to understand why we sometimes picture the mind as radically distinct from the body. This is especially so if we compare human cerebral cortex capabilities with human autonomic responses to temperature changes or to blood sugar levels. Our cortical functioning, in contrast to the autonomic, seems far more sophisticated and less constrained by the physical.

But the brain capable of abstractions and intricate creations is itself embodied in physical structures. It is part of a larger cognitive system. Some of that system—such as the cerebral cortex—processes and directs more complex stimuli and responses, and some of it—such as the spinal column—operates more simplistically and automatically, although no less essentially, in fight-or-flight situations. When the cognitive is recognized as fully embodied, more of the body becomes part of the cognitive, as described by Lakoff and Johnson (1999):

> As is the practice in cognitive science, we will use the term cognitive in the richest possible sense, to describe any mental operations and structures that are involved in language, meaning, perception, conceptual systems, and reason. Because our conceptual systems and our reason arise from our bodies, we will also use the term cognitive for aspects of our sensorimotor system that contribute to our abilities to conceptualize and to reason. Since cognitive operations are largely unconscious, the term cognitive unconscious accurately describes all unconscious mental operations concerned with conceptual systems, meaning, inference, and language.
>
> (p. 12)

We even carry neurons in our stomachs, in our enteric systems, that give us what we colloquially call our "gut" responses (Furness, 2006; Gershon, 1998). And the complexity with which our hormonal and immune systems work (or malfunction), as if with wills of their own, further breaks down the mind–body split. Our bodies are characterized at all levels by intelligible organization as well as by physiological processes—cerebral, endocrinal, digestive, cardiopulmonary, limbic, and so on—that function outside our immediate awareness.

Although most human relations professionals are similarly willing to concede that the mind and body are not separated, the concept of bodymind integration is difficult to retain, in mind or in body, because it can go against our lived experience, especially when we are activated, anxious, or otherwise stirred up.

The challenge of working with people—in our roles as teachers, organizational consultants, counselors, and psychotherapists—is that we have to address these twin problems (of the body and of the mind) as one. But an additional factor complicates our understanding of the integrated bodymind: Each of us is not just one. Our bodies do not stop at our skin. We are linked physiologically with others through resonance, anxiety, reactivity, mirroring neurons, and exchanged cognitions.

This is nowhere more evident, if barely conscious, than when we work in groups.

Living in reaction to others: an organizational engagement

I once worked as a consultant with a large, private-sector corporation that delivered services primarily to local, state, and federal government. Among the innovations introduced by the organization's leadership were cross-functional work teams composed of individuals from finance, marketing, operations, information systems, and so on. The stated purpose of these teams was to improve communications among these functional areas. Their most unusual feature, however, was that no single person was in charge. Leadership was a distributed, shared function. This was a challenge to some group members because the team model contrasted sharply with the hierarchical style of the organization. Other team members welcomed the increased communication and chance to contribute as peers.

One afternoon, I was told to attend one of these team meetings the next day. I was told to "find out what's going on," a typically fuzzy and pressured assignment. I knew that members of this team had been involved in a project for one of the company's most important accounts, a project that had been badly mishandled. At the order of senior management, the team had been meeting to find out what had gone wrong, ostensibly to prevent similar mistakes in the future. But they had reached an impasse. I knew I was supposed to unblock things. When I showed up the next day, I felt both anxious and angry to be doing something that seemed destined to fail.

Several team members immediately demanded to know why I was there, called me a spy for management, and claimed they could not speak freely with me present. My efforts to appear neutral and interested in helping were met with scorn. However, a couple of other team members said they were glad I was there because they thought the group was in trouble. Since those two were well liked and trusted by the group, I was allowed to stay, but my role had been subtly redefined from observer to facilitator. I went along without thinking, hoping to win the group's trust. Before I could get my bearings, though, members jumped into discussion. More accurately, they began fighting, blaming, and making excuses, with a great deal of hostility. When I eventually voiced my observation, several group members turned on me angrily. I backed off.

The tension in the room was explosive. I had trouble breathing and maintaining my composure. I wanted to retaliate. It took me a moment to recover my ability to think about the group's process. And when I came back to my senses—back into my own skin—I realized with alarm that this group was intently looking for someone who could be fired. By the time I had scanned the group to identify the most likely victim—studiously avoiding the group's hostility toward me—I saw with sickening horror that the person had already been picked. She was a younger African-American woman (in a largely white company), not as educated as the others nor as experienced in business or politics. She was floundering and defensive, inadvertently implicating herself as, one by one, the group members came to an unspoken agreement to single her out. At first I could not believe it was happening because I knew many of the people on the team. I knew they usually worked with a spirit of camaraderie. Why were they doing this? I voiced the question so timidly that no one heard me. One of the team members who had originally objected to my being there even gave me a warm smile as if to say we were finally working together. And once the team had "decided" (without explicit words) who would take the fall, they worked quickly and efficiently to bring the meeting to a close. I sat there stunned and appalled as people filed out of the room, many of them thanking me for being there, the targeted woman having fled at the first chance. The next day I was even congratulated by senior management for a job well done.

This team had picked up an unspoken directive from senior management to find a scapegoat whose sacrifice would mollify the angry client. No one in the group knew this consciously, but the organizational anxiety was such that everyone understood, in their bodies, what to do. And I was swept up, too, especially in my attempts to avoid the group's ire. I do not believe anyone intended to be hurtful. In fact, each of us seemed ashamed when we later crossed paths. To this day, I still feel a piercing shame. Yet on that day, we had not been able to access our thinking or tolerate the buildup of tension and fear. Instead, we reacted ruthlessly and without thought—without sufficient skepticism or compassion.

That kind of pain associated with human relations can sometimes lead us to withdraw from contact, as I will discuss further in a subsequent chapter. Arendt (1945/1978) described such recoiling as the "shame at being a human" (p. 235). But the gut impulse to retreat ignores the fact that finally we are not separable from one another. Although withdrawal may offer short-term relief from interpersonal pain, we must, in the long run, reengage and negotiate lives that are fundamentally interdependent.

The shared bodymind

Berne's earliest papers on human behavior were republished after his death in the edited volume titled *Intuition and Ego States* (Berne, 1977c). They reveal how he was, early in his psychiatric career, particularly interested in our capacity for intuition, that is, for processing stimuli outside awareness. He wrote:

> Biologically, intuition may be related to primitive cognitive processes in lower animals (Darwin, 1886; Krogh, 1948; Wiener, 1948b). Phylogenetically, it preceded verbal knowledge and communication (Sturtevant, 1947, p. 48), and ontogenetically as well (Deutsch, 1944;

Shilder, 1942, p. 247). Psychologically, it is important because it is related to problems of group behavior and their limiting case, "what happens between two people," which is the nuclear problem of everyday living and of psychotherapeutic technique.

(Berne, 1952/1977a, p. 42)

Berne devoted his professional life to exploring this "nuclear problem of everyday living," which he saw most irreducibly in terms of transactions or "what happens between two people." The centrality of relational transactions is such that our cognitive processes are most fundamentally oriented toward fellow human beings. In his writings on intuition, Berne (1952/1977a) captured the sense of urgency with which we observe and rapidly assess one another:

Human beings ... behave at all times as though they were continually and quickly making very subtle judgments about their fellowmen without being aware that they are doing so; or if they are aware of what they are doing, without being aware of how they do it.

(p. 35)

This kind of instinctive responsiveness to others can be most readily seen in human families. Based on observations of such interdependent, mutually cued behavior, Bowen (1978/1994) developed his conception of families as natural living systems or organisms. Like Berne, Bowen was a pre-World War II, Freudian-trained psychiatrist who went on to develop his own theory of human functioning, specifically a theory of families. Bowen believed that individual behavior could not be understood in isolation from the family emotional systems into which individuals are born and of which they remain an inextricable part throughout life. He saw individual behavior as a function of the flow of energy or anxiety-based responsiveness of system members to one another as well as to outside threats and opportunities. He conceived of that groupish, less-differentiated component of individual behavior as operating in tension with individual initiative and the capacity for using our minds to override automatic behaviors. Moreover, Bowen considered such tendencies toward undifferentiation and differentiation to have evolved over time in the human species as a function of survival. That is, we have evolved as a species whose lives have depended on our functioning in family groups, in emotionally interlinked and responsive systems.

According to Lewis, Amini, and Lannon (2001), our limbic systems are a component of the cognitive-physiological unconscious that we effectively share and regulate through mutual cuing. Over millennia, "mammals developed a capacity we call limbic resonance—a symphony of mutual exchange and internal adaptation whereby two mammals become attuned to each other's inner states" (p. 63). This tendency toward attunement is so compelling that "the limbic activity of those around us draws our emotions into almost immediate congruence" (p. 64). Therefore, "because human physiology is (at least in part) an open-loop arrangement, an individual does not direct all of his [or her] own functions" (p. 85).

Our psychophysiology is, in part, functionally shared with others of our kind, particularly when it comes to human growth and development. Our struggles to learn and adapt are carried out in the intimate, interdependent company of other divided selves. The paradigmatic example of this shared bodymind is the mother–infant pair. As Winnicott (1952/1975) famously remarked, "There is no such thing as a baby ... If you show me a baby you certainly show me also someone caring for the baby ... the unit is not the individual, the unit is an environment-individual set-up" (p. 99).

Although the mother and infant are each capable of independently handling basic bodily processes (breathing, digesting, sensing, patterning, willing, etc.), each is also involved in a process of reciprocal regulation. This is the "environment-individual set-up" (Winnicott, 1952/1975, p. 99). Infants are dependent on their caretakers for food, shelter, and warmth. But human adult caretakers are equally responsive to cries, movements, and other signs that evoke instinctive parental response sequences: rushing to avert danger, coddling, chiding, barking orders, smiling, cooing, and so on. Such urgent, commonly observed responsiveness has likely evolved over time. True, some mothers (and fathers) are not psychophysiologically activated by their offspring, and some babies manage to survive with minimal adult care or involvement. But by and large, human adult–infant pairs are reciprocally regulating, or attached, as if their lives depended on it.

In an overview of the then-emerging developmental neuroscience research, Schore (2005) defined attachment as "the dyadic regulation of emotion" between mother and infant (p. 206), akin to the concept of limbic resonance. However, he did not mean just attunement or mutual pleasure but also patterns of reciprocal, often asymmetric engagement. Moreover, he added, *"Throughout the life span*, attachment is a primary mechanism for the regulation of biologic synchronicity within and between organisms" (p. 207, italics added). In other words, our biological capacity for reciprocally picking up and processing sensory data is a primary characteristic of the shared bodymind at any stage of development.

Such reciprocal, shared processing of sensory data is what Ogden (2005) meant when he wrote, "It takes at least two people to think" (p. 64). In this view, thinking is not just an autonomous capacity of what Berne would call an Adult ego state. Rather, thinking can also be seen as an acquired or developed capacity requiring the active presence of others. It can be a capacity we must continually learn anew with others as we grow in life. In Ogden's conception, thinking is not an endpoint in the development of adult human beings. It is an ongoing, interdependent process.

Thinking can thus be understood far more broadly as the organization of both conscious and nonconscious sensations and cognitions derived from multiple levels of experience. Thinking, in this broader view, is how humans process those disparate life experiences, making them meaningful and useful. And depending on the mental and physiological difficulty of the experiences, it may take at least two people for thinking of that kind to occur. It may take at least two people for reciprocal regulation in the bodymind to organize itself into usable cognitions. For this reason, Sullivan (1953) went so far as to characterize "unique individuality" as a "delusion" (p. 140) given what he saw as our physiological and cultural interdependence throughout the life span.

Yet many of us carry the idea that we are fully individuated. We sometimes believe that our sense of reality is not constructed and verified in relationship with others; we sometimes forget that our physiological responsiveness is also a function of the emotional tumult in those around us. Conversely, some of us also live as if we had no need to negotiate our separateness.

In my own experience as a teacher, I have found it particularly challenging to perform the overt task of conveying information while supporting the often unspoken aspiration that learning be a transformational experience, as Bollas (1987) so beautifully described that desire for maturing. I feel often caught between student demands that I do all of the work—not wholly unreasonable given that I am the one with some experience and information to impart—and the anguish they encounter in having to take up learning and personal responsibility in order to grow.

Maturational impasses: an example from teaching

I was standing in front of a class of 25 adult learners. This was our sixth three-hour evening meeting, with nine more weeks to go. We had a lot to cover. Several people in the front row and in the back were sitting sullenly, passively, daring me to give them something that would matter. Several had not yet arrived. I had already been receiving their assignments late (always with credible excuses), poorly done (with surprised reactions to my corrections), and executed with little spirit. The message was, "It's your job to get us through this class with a good grade." Their attitude mirrored that of the university department where I had been hired as an adjunct faculty member. Standing before the class, I felt heat rising up my shoulders and flooding my face. I could not remember what I planned to say. I wanted them to feel the excitement I felt for my material. I wanted them to want to learn. I was not yet aware how I hated them, the experience, and my discomfort. Yet even with the likelihood of no change, I could not just walk out the door. I needed the job.

Over the course of two years, I obsessed about the reactions of certain students. I felt ashamed that I could not get the class sessions to work. I dreaded the evenings I would teach. I pushed through my lectures, my unsuccessful efforts at experiential learning, and my failed attempts to elicit participation. I fretted about the passing grades I felt obliged to give. I felt caught in an agreement not to challenge the status quo. Yet I also knew that courses like mine allowed these adults—mostly older, minority women—to get better-paying jobs. I sensed the quality of a game but could not name, much less, master it. I could not recover my ability to question the overt messages of the university department, my immediate supervisor, the students, or society. I could not tolerate the complex lives that stared back at me from faces that affected boredom and invulnerability. I could not link my material to what mattered in their lives. I could not connect the process of learning to something that felt like hope or excitement.

This situation did not improve for the two years it took me to find another job with another college. Even years later, I am haunted by this failure. Not even subsequent teaching successes have helped me come to terms with a surfeit of mixed and crossed messages—coming at me in a profusion I could not sort through. Later successes have not resolved the upset I felt then and now, which I currently see as an embodiment of the troubles we face, as a community, in learning how to teach, make curricula relevant, help people acquire skills and opportunities, and engender a sense of ownership for the ability to learn and grow.

With hindsight, I see this impasse—my inability to facilitate the development of my students (or of myself as a teacher)—as a collapse of my ability to think independently or creatively or to tolerate the emotional data of the experience. Looking back, I wonder what would have happened if I had identified the stance I wanted to take in a situation I could not change and one within which I chose to remain. In a classroom, as in any group setting, I participate in a complex shared bodymind. There is the psychophysiological process of the class as a whole. There are processes within subgroups, as there are within the individuals in the class. There is also the cognitive-emotional process of the organizational system of which the class is but one component. And beyond that, there are political and social systems that have stakes in the outcomes of the class, outcomes that do not always serve the individuals or the community.

Could I have remained within that complex psychophysiological system as an individual—able to think my own thoughts, able to question the unquestioned assumptions, and willing to suffer the anguish of those whose lives intersected for that span of 15 weeks? Could I have adopted a stance of separateness while remaining part of the system, not just because I needed the job but because I recognized I could do nothing to solve the problem by taking the fast road out of there?

I believe this anguished tension can attend our work at any stage of professional development, especially when we continue opening ourselves to more of our world and the people in it. Fortunately, our theories offer us a structure for thinking about and being with such mounting complications.

Transactions inside and out

The conditions of the bodymind, particularly when reciprocated by another, can be staggeringly complex. Freud's effort was to find a manageable way to think about and respond to that complexity. To a considerable extent, he left a legacy that can offer bracing clarity in terms of thinking of the unitary bodymind. Berne himself learned from those early theoretical and practical efforts. However, in response to psychoanalytic tendencies toward arcane theorizing, Berne attempted to make accessible the most important psychoanalytic concepts. In particular, he recognized the importance of analyzing transactional patterns both intrapsychically and out in the world. He proposed an organizing schema for doing so.

Berne's attention to external and internal relatedness was, in part, influenced by object relations theory, particularly the work of Melanie Klein and her immediate followers. Harley (2006) examined that Kleinian influence on Berne's development of the existential positions in transactional analysis.

Object relations theory, as advanced by Klein, Fairbairn, Guntrip, and others, began as an extension of Freud's classical psychoanalytic precepts. It grew out of recognition of the shared bodymind and the need to broaden the usefulness of psychoanalytic thinking and practice to the most disturbed, sometimes psychotic, patients. As such, object relations theory addresses the mind–body split, the confusion of selves that exists between one or more intimates, the interdependent maturational process (as exemplified by the mother–infant pair), and nonconscious communication via the shared bodymind. Although these dynamics may be most apparent in actively psychotic individuals, object relations theory considers such "psychotic" processes to be operative for all human beings to varying degrees.

Winnicott and Bion are arguably the two most important post-Kleinians in terms of grounding their theorizing in the difficulties of actual practice and offering generative ways of thinking about human relational functioning. Ogden (2005) discussed Winnicott's *holding* and Bion's *container–contained*:

> I view Winnicott's holding as an ontological concept that is primarily concerned with being and its relationship to time. Initially the mother safeguards the infant's continuity of being, in part by insulating him [or her] from the "not-me" aspect of time. Maturation entails the infant's gradually internalizing the mother's holding of the continuity of his [or her] being over time and emotional flux.

By contrast, Bion's container-contained is centrally concerned with the processing (dreaming) of thoughts derived from lived emotional experience. The idea of the container-contained addresses the dynamic interaction of predominantly unconscious thoughts (the contained) and the capacity for dreaming and thinking those thoughts (the container).

(p. 93)

For object relations practitioners, the analyst–client relationship is analogous to the mother–infant pair. Both relationships are seen to be functions of the shared bodymind. According to Schore (2005), "The principles of regulation theory that apply to the mother-infant relationship also apply to the clinician-patient relationship" (p. 211). But reciprocal regulation is not considered mutative in and of itself. For the infant/patient to grow and achieve a reasonable measure of autonomy, the mother/analyst must function asymmetrically more than mutually. She must provide structure in the form of holding (in Winnicott's sense) and symbolizing as an outcome of containing (in Bion's sense). Such asymmetric responsibility can be physically and mentally demanding.

As a result of this object relations tradition, many psychodynamic practitioners are accustomed to thinking of treatment and intervention—across a range of human relations activities—as occurring in the dynamic of the shared bodymind. Some transactional analysts also conceive of their work in relational terms. For example, Novellino (2005) cogently characterized relational psychoanalysis and transactional analysis as "addressing the mind's primary need for interpersonal self-regulation" (p. 159). Hargaden and Fenton (2005) claimed that "reciprocal influence and mutual regulation underpins the very notion of intersubjective relatedness" (p. 175). But what might this look like in practice?

By way of example, Eigen (1998) described his work with one patient whose physical tightness he allowed himself to experience, empathetically, for several months before that tightness, known to Eigen from a bodily perspective, began to dissolve into Eigen's joy at seeing his patient. At that point, his patient's nausea symptoms began to lift. Something essential had been transacted bodily more than verbally or even consciously.

Likewise, in my own work as a counselor, teacher, and consultant, I have felt—for extended periods of time and usually with great discomfort—the embodiment of particular clients and students. I have come through those experiences with important understandings in which intellectual speculation is replaced by a tangible knowledge that I could never have imagined or structured without it coming through and from my body. I relate such experiences to my understanding of Bion's concept of containing, which I elaborated in the previous chapter (see also Landaiche, 2005).

One of the male characters in Toni Morrison's (1988) novel *Beloved* explains why his relationship with a particular woman is so important in the context of the life he suffers: "She gather me, man. The pieces I am, she gather them and give them back to me in all the right order. It's good, you know, when you got a woman who is a friend of your mind" (pp. 272–273).

In so befriending the bodymind, our effectiveness as human relations professionals is likewise grounded in our ways of being and in the ways we are experienced by the people with whom we work. The challenge is in allowing the fullness of our capabilities to operate, in an integrated manner, without becoming lost in mental deceptions or intolerance of the body's forces. As Bion noted (1959/1967a), "It is a short step from hatred of the emotions to hatred of life itself" (p. 107).

How do we avoid such "hatred of life" while allowing the fullness of our capabilities?

The simplicity and difficulty of this practice

I have found especially useful Berne's (1968a) contention that:

> One of the most important things in life is to understand reality and to keep changing our images to correspond to it, for it is our images which determine our actions and feelings, and the more accurate they are the easier it will be for us to attain happiness and stay happy in an ever-changing world where happiness depends in large part on other people.
>
> (p. 46)

Often, to understand my professional reality, I write—notes, journal reflections, papers. But that effort at organization can seem futile, even absurd, in the face of the uncontainable complexity that attends every client engagement. That is why I presented here case examples that were not fully resolved. They reflect the reality that I live with daily, even though many times I can also see my effectiveness. I do believe that my struggles to name and organize my experiences—to change my images of them—have substantially helped me to be present as a useful, sentient person.

In any engagement as a consultant, teacher, or counselor, my basic stance is to remain as fully open as possible to the multiple signals I am receiving from others when our contract is for personal growth. Berne (1972) wrote about the Adult listening "to the content of what the patient says, while his Child-Professor listens to the way he [the patient] says it … [The] Adult listens to the information, and his Child listens to the noise" (p. 322).

Although I do not experience my receptivity as embodying two different ego states, Berne did capture the dual strands to which I am continually attending. There is always an overt message, and it is always accompanied by another, more emotional layer of communication. Sometimes those messages are congruent, as when a work group describes a problem it is having and conveys its intention to take up a solution. But sometimes the parallel tracks are wildly divergent, as when a story about a harrowing childhood is accompanied by my feeling deprecated in my role as the listener.

I always give priority to the unworded message that I am picking up in my body. I try to link it if possible to the overt message. Failing that link, however, I simply go with my gut. I try to find some way to describe the senses that have come to my awareness. Naturally, much happens outside that awareness. Yet even if initially fragmentary, that bodily, felt message has repeatedly proven to be the heart of why my clients or students have sought help. And I can often feel it operating the moment they walk in the door, the sensing of which Berne (1977c) described as the first step in an intuitive process.

Although I lead with my receptive bodymind, my skeptical attitude is not far behind. I ask myself immediately, "What is this person really saying to me?" I wonder why I am coming to certain conclusions and whether they are accurate. I also question my clients' or students' conclusions, especially when they seem discordant with the underlying, bodily received message. Although it may sound paradoxical, I experience my skepticism as a corporal relationship to my clients and students and to the material we are working with. That is, I experience a skeptical frame of mind emerging first in the form of an insistence to individuate. This is not always easy for me, because I am more comfortable remaining in an attuned, empathic place with others. Yet I make this physical shift in my sense of self in order to think independently in the midst of intense, although not always obvious, affect. Finding my own mind is always first an activity of my body.

Yet even before I am receptive and insistently reflective, compassion is the primary ground of my practice. I use it to anchor my receptivity because I often find myself reeling, internally, from the nonverbal messages conveyed by the individuals, groups, and students with whom I work. I also use compassion to temper my mental activity, to make sure my skeptical leanings are not merely justifications for rejecting the difficulty I experience in certain professional situations. I use compassion as a guide to recognizing that I, too, have often struggled with the very issues presented to me and to remind myself that I know something in my body of what it is like to sit with certain troubling emotions and still have use of my mind.

Although my theoretical orientation as a practitioner has been shaped in large part by transactional analysis, object relations, group analysis, and Bowen family systems theory, I do not usually think about theory when

I am working. Those ideas have been formative in my professional growth. They certainly inform my thinking when I am not working, as they do in this chapter. But in the midst of my human relations work, I am not thinking theory. I am thinking with my full bodymind, which is structured as a result of everything I have already lived and learned. I am opening myself, often with some painful difficulty, not just to the other person but to the capacity we all have, as human beings, to organize our experiences and to respond in a manner that is consistent with the reality before us, which in each professional situation is the client system seeking our help.

Skepticism and compassion are obviously not panaceas. They do not necessarily remake the world. Yet they do allow us to reregulate the workings of the bodymind—unitary and shared. In so doing, we acquire the means to make considered choices on behalf of ourselves and those with whom we make common cause.

I conclude in that spirit and with what African-born, French psychoanalyst Franz Fanon (1952/1967) called his "final prayer": "O my body, make of me always a man who questions!" (p. 232).

3
LEARNING AND HATING IN GROUPS

Even after years of studying groups and leading them, I found myself mostly stuck, not growing, still deeply troubled. So I began to examine my own membership, my own patterns, pursuits, and reactivity, even when in a leadership role. I used myself as the research subject—again adopting the attitude of *engaged research*—to learn about myself, certainly, but also to discover more about how my groups themselves got stuck. What is the nature of this collective I am trying to work and thrive within?

This chapter is not about group leadership or training but about being in (and sometimes avoiding) groups, about the learning that can occur there as well as how groups can interfere, sometimes hatefully. Although these aspects of groups and learning certainly bear on leadership and training, I want to address the much more vulnerable and disorganizing position of membership, especially the considerable effort it can take to mature among others without being harmed or stunted.

As in previous chapters, I am using the term *group* to refer to teams, organizations, professional or other communities, classrooms of students, "groups of all sorts and sizes from nations to psychotherapy groups" (Berne, 1963, p. 4), even families. I like Berne's inclusive term "social aggregation" (p. 54) with its "external boundary" (p. 54) distinguishing members from nonmembers. I will also explore the theme of group culture, in particular that of the transactional analysis community as I have observed its effect on my learning in groups.

Berne's group legacy

Group dynamics and group work played a key role in the development of Berne's (1961b) thinking and his eventual formulation of transactional analysis, which he characterized as an "indigenous approach derived from the group situation itself" (p. 165). At least ten years before his two best-known books on the topic (Berne, 1963, 1966), Berne's first publication on group therapy principles appeared in the *Indian Journal of Neurology & Psychiatry* (Berne, 1953). He went on to address the dynamics of a peer-initiated personal therapy group (Berne, 1954), the technical considerations of group attendance (Berne, 1955), transactional analysis as a form of group therapy (Berne, 1958b), cross-cultural factors (Berne, 1958a, 1961a), contrasts between psychoanalytic and dynamic group therapy (Berne, 1960b), the training of group therapists (Berne, 1962), and the unique opportunities for learning and growth in heterogeneous meetings of the kind he called "staff-patient staff conferences" (Berne, 1968b). A sample of Berne's actual method of working in groups was also published as a verbatim transcript (Berne, 1970/1977b) shortly before his death.

Yet Berne's written legacy does not apparently offer a complete theory of groups, at least not in comparison to his highly articulated theory of individual psychology in the context of significant interpersonal transactions. Nonetheless, his writing on groups and organizations, his ways of working, and the professional culture he established seem to have been especially generative to the thinking, practice, and writing of others in transactional analysis over the past 50 years.

I want to acknowledge some part of that ongoing effort. For example, in two issues of the *Transactional Analysis Journal* devoted to groups (Cornell & Bonds-White, 2003; Stuthridge, 2013) and in other publications, certain authors have attended to the use of groups for counseling and psychotherapy (Boholst, 2003; Caizzi & Giacometto, 2008; Clarkson, 1991; Cory & Page, 1978; Erskine, 2013; Hargaden, 2013; Hargaden & Sills, 2002a; McQuillin & Welford, 2013; Misel, 1975; O'Hearne, 1977; Steele & Porter-Steele, 2003; Steiner & Cassidy, 1969; Tangolo, 2015; Tangolo & Massi, 2018; Tudor, 1999, 2013). Others have explored the application of group treatment to special populations (Arnold & Simpson, 1975; Capoferri, 2014; Cassoni & Filippi, 2013; Kinoy, 1985; Sinclair-Brown, 1982; Spence, 1974; Thomson, 1974; Tudor, 1991).

Organizational systems and consulting have been a focal point for a different set of practitioners (Altorfer, 1977; Balling, 2005; Blakeney, 1978a, 1978b, 1983; Brown, 1974; Cardon, 1993; de Graaf, 2013; Hay, 2000; Jacobs, 1991; Krausz, 1986, 1996; Kreyenberg, 2005; Mazzetti, 2012; Mountain & Davidson, 2005; Nuttall, 2000; Nykodym, 1978; Nykodym, Freedman, Simonetti, Nielsen, & Battles, 1995; Petriglieri & Wood, 2003; Poindexter, 1975; van Beekum, 2012, 2013, 2015), with some of their contributions gathered in themed issues of the *Transactional Analysis Journal* dedicated to organizational work (Groder, 1975; van Poelje, 2005).

Some authors have looked more explicitly at group dynamics as they show up in the broader spheres of community and politics (Cornell, 2016, 2018; Deaconu, 2013; Samuels, 2016).

In the areas of training and education, the group's essential role in the learning process has been elucidated by Haimowitz (1975), Kuechler and Andrews (1996), Clarke (1981), Crespelle (1988), Ranci (2002), Bonds-White (2003), Hawkes (2003), Newton (2003), and Landaiche (2013).

Allen and Hammond (2003), Noce (1978), and Robinson (2003) have written about the interplay between institutional factors and treatment groups. Still others have looked at the dynamics of groups irrespective of practice contexts (Bonds-White & Cornell, 2002; Campos, 1971; Gurowitz, 1975; N. L. James, 1994; Krausz, 2013; Landaiche, 2012; Lee, 2014; Micholt, 1992; Sills, 2003; van Beekum & Laverty, 2007; Woods, 2007) whereas some have examined the contrasts and complementarity of different theoretical and practice approaches (Dalal, 2016; Kapur & Miller, 1987; Peck, 1978; Shaskan & Moran, 1986).

Personal perspectives have also been welcomed as part of this ongoing dialogue and exploration (Solomon, 2010; Wells, 2002).

This roll call of names and applications attests to an exceptional history. Along so many paths, group life has played a key role in training, learning, and practice within the international transactional analysis community. One might say that Berne's failure to leave a fully articulated theory of groups, in fact, contributed to a culture of freedom to be curious about groups and to use and think about them in myriad ways. As a result, I believe a strongly coherent attitude toward group life is passed along implicitly, via cultural means, to succeeding generations of transactional analysis practitioners. I imagine one could trace those same cultural influences in the object relations and family systems traditions as well.

Culture in transactional analysis

Well before Berne formally designated transactional analysis as a theory, the theme of culture, like that of groups, appeared frequently in his writings. In addition to his interest in the effects of cross-cultural factors on group therapy (cited earlier), Berne visited and wrote about psychiatric practices in Syria (Bernstein, 1939), the Fiji Islands (Berne, 1959a, 1959c), and Tahiti (Berne, 1960a). He also visited and compared practices and conditions in psychiatric hospitals in the Philippines, China, Singapore, Malaysia, Sri Lanka, India, and Turkey (Berne, 1949), eventually writing about the field of comparative psychiatry (Berne, 1956). He also considered the psychiatric uses of mythology and folklore (Berne, 1959b) as well as the role of culture in the case of a Filipino man who had murdered five friends (Berne, 1950).

Berne (1963) wrote explicitly of culture in groups as "the material, intellectual and social influences which regulate the group work, including the technical culture, the group etiquette and the group character" (p. 239). He considered culture a key component—along with the "the Constitution ... [and] the laws" (p. 96)—of the group's canon, "a regulating force" that gives "form to the group cohesion" (p. 239). He observed that "the culture influences almost everything that happens in a social aggregation" (p. 110). Yet in his last book, Berne (1972, p. 323) was dismissive of culture, claiming, for example, that it had little to do with individuals' scripts, a position that Mazzetti (2010) pointedly argued against.

Within the transactional analysis community, the significance of culture was explored early on by Roberts (1975), White and White (1975), and in a special issue of the *Transactional Analysis Journal* (J. James, 1983). Drego (1996) later elaborated her idea of culture as a societal Parent with oppressive aspects that had regenerative potential. Within the past 17 years, other authors have also explored the culture of the transactional analysis community itself, particularly the influence of that culture on practitioners' ways of working (Campos, 2012; Cox, 2007; Erskine, 2009; Grant, 2004; Mazzetti, 2010; Newton, 2003, 2011b; Noriega, 2010; Oates, 2010; Robinson, 2003; Tudor, 2002, 2009).

As Newton (2003) wrote:

> The transactional analysis community has a common premise in our "statement of faith": "I'm OK, You're OK," a philosophical base reflecting our ultimate values. Teaching transactional analysis is also "doing" TA. The question is, how can our training demonstrate "I'm OK, You're OK"? How do we pass on our culture?
>
> (p. 321)

This attention to culture and its powerful transgenerational effects fits, I believe, with the important role that anthropologist Mary Douglas gave to institutions. She developed this thesis in her book *How Institutions Think* (1986), setting forth a view that I find quite consistent with Berne's writings about the centrality of the group's, organization's, or community's canon.

By-products, climate, action

In thinking about group or institutional life, I use the word *culture* in three ways. First, I mean a set of tools or schemas for living together, which can include certain foods, customs, rituals (and yes, games), rules, wisdom handed down, objects of art and music, literary works, fashions, and other cultural products. Additionally, for me, culture refers to an environment, an atmosphere within which life is fostered over generations, one that develops a particular feel and smell nearly impossible to put into words. Finally, I use culture as a verb for the intentional shaping of life, as in husbandry or farming (see Barrow, 2011).

I consider culture a collective effort and organization intended to further life. Damasio (2010) wrote about the evolution of what he called "sociocultural homeostasis" (p. 294), the idea that culture—that is, "justice systems, economic and political organizations, the arts, medicine, and technology" (p. 26)—develops in a biological and social context to promote the survival of living organisms, in this case human beings. Berne (1963) alluded to this essential aspect of group life when he wrote that "preserving the group culture takes precedence over the group activity" or primary task (p. 20).

So, when I write of culture in the transactional analysis community, I refer to concrete objects and procedures left by previous generations; I allude to the unique feel of being in that community; and I point to the efforts of community leaders and teachers to facilitate continued growth, especially when nurturing the next generation.

"Toward hard therapy and crispness": a Martian cultural perspective

The transactional analysis community structures time and work in a way that seems to demand group participation: training, therapy, conferences, work groups, peer learning, and so on. What cultural attributes might affect this community's approach to group work and learning, as Tudor (2009) outlined the attributes that inform a transactional analysis approach to training?

One significant aspect of the culture, in my view, is Berne's aggressive style of writing, which inspired this section's header: "Toward hard therapy and crispness" (Berne, 1966, p. 104). This was a blunt declaration of separatism that distinguished his hard, crisp approach from the more traditional or entrenched "soft" (i.e., ineffective) therapies. Berne's use of words can be evocative and illuminating, even when his meaning is not precise. Moreover, his style carries an implicit permission to provoke, play, disagree, and make trouble. In fact, the transactional analysis community seems to attract people with independent, sometimes rebellious, turns of mind. There is permission to think and freedom to discover and speak the wisdom one already knows or is capable of embodying, no matter how schooled one is in transactional analysis itself.

Berne (1968b) advocated

> the abolition of professional categories during the [staff-patient staff] conference [as] a license for everyone to think without artificial restrictions: nurses can think like doctors if they wish, doctors can think like nurses, psychologists can think like social workers, and so on.
>
> (p. 289)

In groups, this kind of license can be refreshingly alive and freeing; it allows greater diversity of opinion and contribution and thus greater potential for learning.

Transactional analysis theory provides a frame of reference—a shared language—that remains open to other frames of reference. I think this offers a helpful commonality while still allowing the emergence of new knowledge and unfamiliar forms of expression in groups.

Another aspect of the transactional analysis community that stands out for me is its interdisciplinarity (Landaiche, 2010), the way its theory supports work in a variety of settings and fields of application (i.e., education, counseling, organizations, and psychotherapy). This seems to foster a more complex understanding of the human condition, especially as lived in groups. For although in practice there can be intolerance among the fields, there seems at least a greater chance for professional dialogue and exchange compared to the far larger lack of awareness of such interdisciplinarity in other professional communities of which I have been a part.

There is in transactional analysis a tradition of doing personal work in groups, which normalizes the troubles many of us face and models a process for moving forward. Transactional analysis supports the belief that individuals have the capacity to grow, to outgrow problematic behavior, while framing that potential within realistic limits. In groups, this can offer a sense of hope without promoting an illusion that proves unsupportable in the larger world.

Transactional analysis group environments seem greatly eased by a culture of offering strokes, holding "OK–OK" positions, reducing gamey behavior, and striving for more intimate, honest contact. There is an effort to recognize and meet needs for time structuring. There is a strong "gift culture" (Newton, 2011b) that supports community service and generosity along with a sense of social responsibility—key aspects of productive and ethical group life.

These aspects of independence, interdisciplinarity, intimacy, thoughtfulness, and generosity converged for me most keenly when I co-led a workshop at an international transactional analysis conference. I made what felt like an embarrassing mess of my presentation by allowing myself to go "off script"— neither following my prepared notes nor heeding the injunction to speak only when rehearsed. Although I was mortified, the group did not shame me. Others took my spontaneity as an opportunity for lively exchange and learning in the group. Although in that setting I was one of two designated teachers, my own deep learning arose largely from the conference attendees' collective abilities and willingness to make something of my disorganization, as they did by giving me feedback, by allowing themselves to be touched, and then returning what they made of it to me and to the larger group. It was an extraordinary experience of risk and consolidation, a microcosm of the growth that community can cultivate.

Some glitches

Along with these gifts of the transactional analysis culture, are there also aspects that interfere with learning and growth, especially in groups?

When I have observed splitting and divisiveness among the different fields of transactional analysis application, certain "consensual imagoes" (Newton, 2003, pp. 322-323) appear to be reinforced, supporting, for example, infantilization of trainees or intellectualization in relation to our complex, fraught organizational processes. Sometimes we pathologize difficult human emotional experiences at a clinical remove. Or we fly into action prior to sitting with and fully grasping the emotional and cognitive difficulties that clients and students bring with them. I believe these more narrow views of the human condition limit the potential of groups to do their work and thus constrict members whose own learning difficulties need a broader scope.

There seems also to be a downside to the straightforward clarity of transactional analysis theory, its dialect, and its emphasis on thinking and rationality. After all, groups can be intensely irrational, unthinking, contorting places. How to find words to speak of this? Berne attempted to convey this more disturbing aspect of human life with his concepts of "mortido ... the energy of the death instinct" (Berne, 1947, p. 305), the "demon ... nonadaptive impulsive behavior" (Berne, 1972, p. 443; see also Novellino, 2010), and the nonconscious, compelling pattern for script he called *protocol* (Berne, 1963, 1966, 1972; see also Cornell & Landaiche, 2006). Yet it seems hard for us as a community to sustain our attention to these darker aspects long enough to shape them into meaning.

Although I spoke earlier about the freedoms that may arise from the lack of a fuller transactional analysis theory of groups, I wonder if that lack also makes it difficult for community members to be aware of and find words for significant group processes and so to respond differently within them.

Over 40 years ago, Schanuel (1976) cautioned against what she saw as an emerging insularity, "an in-group feeling among TA enthusiasts ... [that is] strengthened by the use of a special language, increasing complexity of theory, and a very disturbing emergence of a belief that TA can do everything" (p. 316). She pleaded, "Let us not in-group ourselves out" (p. 317). This speaks to the other side of the "respected marginality" that Petriglieri (2010) described in his keynote speech at an ITAA conference in Montreal, where he referenced the transactional analysis community's respectably marginal relationship with the larger fields of education, organizational consulting, counseling, and psychotherapy.

Perhaps every cultural strength has its problematic underside.

My group psychology

Using myself as a research subject within the transactional analysis culture, I will share some of my own history and psychological makeup by way of context.

As I wrote about being the eldest of nine siblings, my script role was to overfunction: make peace, pick up the slack, keep an eye on everybody and everything, clean up the mess, take charge, be responsible, ad infinitum —a torrent of injunctions, all varying the same tired theme.

Undergirding my scripty behavior, I see a primary preoccupation with survival, with avoiding, at all cost, rejection and social pain, which I will elaborate in the next chapter (see also Landaiche, 2009). Some of my worry may be due to having been disrespectfully, hatefully marginalized at times—the worst being when I was relentlessly bullied for four years in boarding high school. But I do not think my personal history fully explains my expectation of maltreatment.

In simple terms, my primary approach to groups is to withdraw, to get out of harm's way.

Although I first thought of withdrawal as cutting off or sulking silently, I now see that I can withdraw even when it looks like I am actively participating. This happens when, in the midst of action, I lose track of myself and what I am trying to learn or get from the group, when I become involved in controlling or overfunctioning for the group, when I indulge fantasies of dramatic withdrawal rather than think about what is frustrating and what I can do about it. These subtle, sometimes disguised, forms constitute a withdrawal from the group in which I also withdraw from and abandon my own growth and self.

Undoubtedly, there are times when one must run for cover. For along with what Rogers (1967) characterized as "the constructive potency of ... group experiences" (p. 263), he also acknowledged that "in some ways this experience [in groups] may do damage to individuals" (p. 263). Yet in my more recent groups, I cannot say that I am typically in any real danger; I just live as if I were, delusionally. As a consequence, I do poorly, becoming more anxious, paranoid, and depressed, and thus less resilient, maturing, or learning. I have also observed that when I can overcome my miserable gut alarm, when I can make contact with my groups, these same problems reverse themselves substantially. And when I participate as an individual, instead of as a role (e.g., overfunctioner, peacemaker), it seems my groups also do better.

For these reasons, I have come to see the problem of my membership in groups as one of withdrawal. And without reversing the habit, I obtain no benefit from membership. I get only the slow, inexorable damage that comes from the illusory safety of disconnection.

In this sense, I am writing to and for those whose last wish on Earth is group participation and who also recognize the costs of gratifying that wish as a way of life.

A counterphobic pursuit

One might say, given my sibling position, that my interest in groups and group leadership was as much trained into me, from an early age, as it was actively pursued from an emergent, developing interest. I think that is true but in reverse. That is, my keen interest in group life began counterphobically. Early experiences had left me intensely shy and mistrustful; I hated groups. Yet in my early thirties, for reasons that are still unclear to me, I felt compelled to go toward what I wanted to flee from, to force myself into some contact with groups of people. I had this sense, still vague then, that if I did not figure out how to be with others, in community, part of an extended family, I would not become myself. I felt that becoming was as much a product of communal processes as of individual effort, character, or genetics, and that I had no integrity except in vigorous interaction with others—family, circles of friends, colleagues, fellow citizens.

As Berne (1963) noted,

> One purpose of forming, joining, and adjusting to groups is to prevent biologic, psychological and also moral deterioration. Few people are able to "recharge their own batteries," lift themselves up by their own psychological bootstraps, and keep their own morals trimmed without outside assistance.
>
> (p. 159)

Yet, what if we have a hard time making use of that outside assistance?

On hating, not learning

Hatred is one key aspect of group life that we must often face first. As Bion (1959/1969) wrote:

> It is clear that when a group forms the individuals forming it hope to achieve some satisfaction from it. It is also clear that the first thing they are aware of is a sense of frustration produced by the presence of the group of which they are members.
>
> (p. 53)

I know well the frustration Bion speaks of, and in the face of it, satisfaction often eludes me. However, rather than talking generally about the hatefulness of groups, I will describe what I hate about them and how that leads to my problematic membership, even when the culture is largely facilitative.

Although hate is not too strong a word, it does not describe the whole picture. After all, I do like being within a group's aliveness—the variety and dynamism, the not having to work as hard (being able to sit back). Life in a group can seem less an effort, less a fleeting and fragile phenomenon.

But I also fear and detest the flip side of that aliveness—the boredom of the group's resolute avoidance, the tensions, the threat of being killed or humiliated (social death), the passivity, the entrenchments, the slowness of deliberation (compared to the quickness and surety of my own mind), the magnification of meanness.

I can feel extreme, panicky discomfort in joining a group when I feel I do not belong, when I feel too different, or when I buck the group canon—for example, when changing seat locations in a group where seats are unofficially claimed or declining food and drink that are part of a group's core rituals.

Sometimes I experience fruitlessness. For example, at the mental health clinic where I work, in our weekly presentations of highly difficult cases, I can get pushy trying to speak my view—bullying the group without actually offering a lead. Or I cannot speak for myself because the group's input keeps interrupting my train of thought. At loggerheads, I fall silent, sullen. How often in groups do we find ourselves facing this kind of sinkhole?

Frustration, uncertainty, contagion, threat—at such times, groups seem hardly worth the effort. Yet away from groups, I can forget what they are like. I romanticize them, an idealization that is rudely crushed when I return to actual contact.

This chapter is thus the more realistic story I tell myself to bear being part of groups.

The part I play

The kind of hating I do in groups, when magnified among multiple members, may account for some of the awfulness groups can realize.

For my part, I can easily disown groups, blame them as though I had no role in their behavior. As Bion (1959/1969) wrote, "The individual cannot help being a member of a group even if his membership ... consists in behaving in such a way as to give reality to the idea that he does not belong to a group at all" (p. 131).

That describes well my kind of disavowal.

Consider the intolerance of being wrong: Clearly, I affect the group when I cannot relinquish what I think on hearing something that contradicts and corrects my view or misperception. I contribute negatively to a group's culture when I have trouble opening my mind to not knowing while in the midst of trying and failing to know. This does not support an environment of free speaking or openness. Intellectually, I concur with Stacey and Griffin (2005), who wrote that

> there is no detached way of understanding organizations [or groups] from the position of the objective observer. Instead, organizations have to be understood in terms of one's own personal experience of participating with others in the co-creation of the patterns of interaction that are the organization.
>
> (p. 2)

Yet emotionally, I still imagine the group as a single, graspable thing and believe, with all my heart, that I see it whole. Even when grudgingly incorporating new data, implying that my previous view was incomplete, even recognizing such revisioning as recurrent (endlessly so), I envision every new perception as the last word. I live no thought of incompleteness, although I may pay it lip service. Since this likely happens for others, too, I merely add to a climate of head butting and fighting over versions of being right.

Group mindlessness

Kreyenberg (2005) wrote that "working with organizations means working with living systems" (p. 300). This is similar to Hargaden and Sills (2002a) characterizing a therapy group as "an interwoven, living, breathing organism" (p. 144). Such aliveness fits my experience of workplaces, families, and groups and may also explain some of the difficulties of community life.

As organisms, groups exhibit sentience. When threatened, for example, they can execute a few basic avoidant or deterrent routines, as can other simple organisms, such as sponges. Bion (1959/1969) called these routines "basic assumptions," primitive, nonconscious conclusions to fight, flee, depend, and so on, all in reaction to intense anxiety. Yet I believe these defensive maneuvers demonstrate a fairly limited intelligence and, when repeated, account for the frustration and danger many of us encounter in groups. Sills (2003) wrote of this repetitive, maladaptive group behavior in terms of "role lock."

Presumably, such thoughtless assumptions and routines have played a role in human survival and, like culture, have been and remain functional in certain circumstances. Many of us may luxuriate in this kind of mindlessness, giving our weary heads a holiday. But group intelligence can seem brainless compared to that of a single human with a thoughtful mind, just as we individuals can likewise function with apparent mindlessness, no better than our groups. Yet like our individual selves, human groups, when faced with complex tasks, need more than archaic groupthink.

When groups can think

For a number of years, I held strongly to the bias that groups cannot think or learn based on what I might consider their most harmful functioning. But over time, I have appreciated the particular kind of intelligence they can demonstrate, which I will explore further in a later chapter. Basically, it appears that when individual group members can think and learn, this clearly stimulates and supports the thinking and learning of other individuals in that group. And that produces a cognitive-like process that is greater than just the sum of those individual intelligences. As van Beekum and Laverty (2007) put it, at those moments "something comes into existence that is not exclusively the product of the individuals involved" (p. 228).

Intelligence emerges, sometimes stunningly, when individuals in a group use their own minds as separate beings. It emerges when a group is containing its diversity, knowledge, and anxiety rather than extruding it. We then see a division of labor—a kind of distributed processing among individuals in the group—wherein a problem is broken down and shared among separate minds. The results of this processing are then regathered to do smarter work, a theme I will return to in later chapters. Douglas (1986) looked at the way institutions consolidate and carry their collective wisdom in relation to highly complex community ethical issues, for example. This wisdom may not always represent the best possible points of view, but it offers a way of thinking, at the group level, that individuals cannot do entirely by themselves.

For individual members to learn and work as a group, it seems they must find and share their own embodied minds. This presence of thoughtfulness and ethical relating can spread through a group as contagiously as anxiety and mob behavior. But clear-headedness, as Bion (1959/1969) noted, is far more difficult because "one is involved in the [group's] emotional situation" (p. 57). To use one's whole mind and body requires disciplined determination and courage, whereas mindlessness we can deliver in our sleep.

In a field of struggling bodyminds

In his paper on "Staff-Patient Staff Conferences," Berne (1968b) wrote that one of the objectives of such conferences is "to stimulate thinking and the organization of thoughts" (p. 286). For me, one value of group life is certainly those moments when everyone is struggling to understand a complex problem, thinking in parallel and speaking with some thoughtfulness or intuition. Yet this is not always so easy. As Milosz (1969/1975) astutely observed—in the gender normative terms of his era—there are times when each of us can be "so ashamed of his own helplessness and ignorance that he considers it appropriate to communicate only what he thinks others will understand" (p. 3). And this would certainly shut down the collaborative thinking process. Yet Milosz also noted that there are times when "somehow we slowly divest ourselves of that shame and begin to speak openly about all the things we do not understand" (pp. 3–4).

This describes well the paradox of human dialogue.

Given my work and life, I find most engaging the challenges of understanding human experience. Even when wanting to hold fast to what I know—shifting uncomfortably in my seat given all the divergent views—I manage somehow to enjoy a sense of growing clarity when others offer their differing perceptions. It expands my framework, accelerates my process, and metabolizes some of what I cannot think alone. I come to appreciate their offerings, in all my twitchy stubbornness.

Personal meaning is different when continually informed, interrupted by, and conversant with this kind of collective emotional and communicative process. Such meaning-making occurs in the sometimes delicate, sometimes vigorous exchanges between and among people, across multiple modes of relating: vocally, tactilely, visually, and so on (see Berne, 1966, pp. 65–71). In listening to one another, with all our senses, we begin to learn the contours of our collective difficulty, perhaps eventually finding a name for it and a place in our larger, shared field of meaning. Schmid and O'Hara (2007) described these as "the extraordinary moments in group life when ... people are ... capable of reconciling complexities that had seemed intractable just minutes before" (p. 98).

We can see such group-enhanced learning in successful brainstorming, in meetings where a clearer vision and more effective planning emerge, in classroom discussions that articulate a phenomenon better than the official text or the teacher's lecture, in treatment groups that encourage the elaboration of certain common life struggles and in which the voicing of multiple perspectives begins to construct a palpable sense of well-being and connection.

Even conflict and disagreement can be clarifying when they give voice to an issue's many sides and make them, eventually, more apprehensible. Sometimes conflict and divergence more realistically represent the state of knowledge than does consensus or apparent clarity.

Stacey (2001) talked more neutrally about this kind of "communicative interaction [that] simultaneously produces both emergent collaboration and novelty, as well as sterile repetition, disruption and destruction" (p. 148). He keeps us in mind of the continual tension to be found in group life between creativity and deadness. My own enthusiasm is similarly tempered. For although I would argue intellectually that certain learning must take place in groups for us to function most effectively in our workplaces, communities, and families, doing so is not always that trouble free or linear a process.

Unsolvable dilemmas

In writing about what happens in groups, Hinshelwood (1987) observed:

> The problem of the individual in the community ... is the problem of securing ... a sense of personal identity. The problem for all individuals is to dare to distinguish themselves from the rest of the meeting. ... [They must] establish clearly the boundary between what they consider is internal and what external
>
> (p. 86)

This underscores a problem I also have observed, for myself and others, in many group situations. For example, a few months ago I attended a meeting of a long-range planning committee of which I am a member. I could not relate to the discussion about markets, products, programs, and delivery mechanisms. I found it pointless. I felt uncomfortable, unsatisfied, unable to speak up, or, more precisely, only wanting to blurt out, to shove the group in my direction (whatever that was). So I kept my mouth shut (too miserably near the fear of speaking). I eventually shifted my attention from the group toward what was squirming in my own mind and body. I then realized I was trying to figure out something that was not part of the overt discussion. I was puzzling over the characteristics of a healthy organization and, specifically, wondering what had once so vibrantly characterized this community service organization for which we were planning. Although I arrived at no answers, my frustration and confusion lessened. My insides found some order. I then could see how the group's discussion, which had seemed so off base, actually contributed to my thinking, a contribution I could only receive when focusing on my own thoughts rather than drifting with or resisting the group. The others, in being contrary, helped me get on track. I was both separate and part of, working on my area of a larger puzzle while they each worked on theirs.

In many groups, however, I often do not find clarity; I just have these questions:

- In the midst of the process and content, how can I remain alert to my own mind and aspirations? Can I figure out what I want and can realistically get from the proceedings?
- In a group that is "wasting its time," what is my responsibility? Is the uselessness an impasse or just the disorganization that often precedes a breakthrough? What must I change in my contribution?
- When I am silent, what is my function for the group: Containing? Listening? Incubating? Mulling (or muddling)? Binding the group's anxiety?
- When "the air is filled with the clamor of analysis and conclusion, would it be entirely useless to admit [I] do not understand?" (Milosz, 1969/1975, p. 4)

I am reminded of what Berne (1966) wrote about the psychotherapy group leader:

In regard to his own development, he should ask himself: "Why am I sitting in this room? Why am I not at home with my children, or skiing, or skin-diving, or playing chess, or whatever else my fancy might dictate? What will this hour contribute to my unfolding?"

(p. 64)

Such unfolding and researching appear to be ceaseless.

Taking self in hand

As an oldest brother, I can step into leadership roles as easily as blinking (talk about role lock). So, I have wanted to learn if focusing on being a more effective group member can inform how I fulfill my leadership positions as teacher/trainer, group facilitator, and senior staff member at the clinic where I work.

What has seemed most salient has been learning to manage my reactive behavior in the midst of group processes. That is, by not trying to coerce anyone else, I relinquish the "reassuringly counterproductive [aim]" (Petriglieri & Wood, 2003, p. 333) of controlling the group.

I try to keep track of my own needs to learn and contribute, which means speaking up and making the most of reality, of the circumstances in that group, at that time. I try to allow curiosity about my uncomfortable emotions; I attempt to bear frustration. With insufficient time for myself (as one of many), I try to sit, maybe miserably, for long periods with very hard-to-solve problems or problems without end, such as those of being human. These could be the efforts of citizenship.

Still to learn

In the past 30 years, I have grown some: I am not as cut off as a family member; I work in an organization, no longer isolated; I participate in professional communities; I am not as aloof. The transactional analysis culture has helped me greatly in this area of maturing, with its by-products (e.g., literature, gatherings, customs), climate of vigorous exchange, and predominant pursuit and nurturing of development. I have been similarly helped by the family systems community, especially here in my hometown, Pittsburgh.

I still have difficulty relinquishing my overfunctioning, oldest-brother role to be an ordinary, active citizen. As I described earlier, I can appear participative when I am actually withdrawing. In regard to being a citizen in my home country, for example, I can voice vigorous criticism and contempt for politics. But I do not take constructive action on a community or political level. Although my complaints may have some validity, I must also own any political dysfunction as partly mine, starting with my abdication of engagement and cooperation.

In the many groups to which I belong, I can see that understanding myself as a member and growing into some maturity as a contributor directly impacts my capacity to provide leadership. I especially cannot provide it when leadership roles offer a way around the problems I encounter in membership, that is, when such roles reinforce my script. Then my leadership is merely a form of overfunctioning, a scapegoated manifestation of the group's undifferentiated ego mass. I just play a part.

As such, as I will discuss more fully in a later chapter, I have come to see leadership as a secondary property or phenomenon, one that emerges from membership. That is, the group member able to manage himself or herself has the potential to become a natural leader who is also able to follow.

Why bother with groups?

In regard to the hopelessness we may feel about group life, and perhaps about life in general, Hinshelwood's (1987) words still ring true for me today:

> Personal despair and despair about the organization go hand in hand. The experience of working in a demoralized organization is not very different from the experience of the patient who has come to the end of the road with his life.
>
> (p. 140)

Yet against this drag of despair that can still haunt my life, I bother with groups and with my discomfort because I am keenly interested in their potential for learning, and for collective struggle and elaboration in relation to certain key problems. In a deliberative democracy, for example, how should we govern, and what is fairness in law and enforcement? In the scientific community, what constitutes evidence in support of understanding how the world works? In a team, how will we accomplish our task? In a seminar on, say, the sacred writings of a religious tradition, how can we articulate the complexity of those often-ancient texts and their unfolding relevance to our lives? Jazz gigs, peer supervision groups, prayer circles, hunting parties, Internet communities of interest: There are countless ways to join in furthering our development across innumerable areas of human endeavor.

One week I was surprised to count my direct involvement in 19 groups, either as leader, follower, or peer. This was not counting the groups that were inactive that week; nor did it include broader community or societal involvement. That same week I also noted how intently I disregarded the extent to which my life was lived always in relation to some group, a fact I still have trouble accepting.

Although I do not believe everyone is a group animal, who, like me, does better when in contact with others, I have come to believe that each of us benefits when we are clearer about our stance toward and the action we want to take in relation to this social feature of human life. To that end, I propose the following basic research questions:

- What do we observe about ourselves and our groups?
- How does our membership affect what we hope to achieve in life and possibly, by extension, what we hope the group and its members will achieve?
- What does our research suggest could be changed in our membership?

We each may approach groups as a danger or pleasure or, prescriptively, as a kind of medicine that, however bitter, will be good for us in the long run. I have come to see them as simply unavoidable. No longer needing to force my encounters with groups, I seek them intentionally, with informed consent, because I grow better as part of their living milieu. I care less whether I do or do not like them. Rather, I passionately want what only they can give to my pursuit of a meaningful life in progress.

4

SOCIAL PAIN DYNAMICS IN HUMAN RELATIONS

As a group consultant, psychotherapist, and teacher, I pay attention to group learning while being particularly interested in extreme avoidant and aggressive behaviors that can be destructive to human relations. These can include isolating, blaming, cutting off, and attacking. The often intense, violent quality of these diverse behaviors can make them difficult to change, especially when accompanied by loss of one's reflective abilities (i.e., Adult ego state functioning).

My interest in these intense, intransigent behaviors led me to study pain, specifically pain dynamics that are activated interpersonally or socially (technically referred to as *social pain* as described later). In that process, I have also become aware of my own pained reactivity in the midst of professional work, which I believe has implications for practice in therapeutic, consultative, and educative settings in which social pain is at play, particularly in groups.

Nonobvious considerations

The fact that human relations can be painful seems indisputable, even obvious. But as I will discuss in a moment, social pain, unlike physical pain, is not always conscious; therefore, it is not always reported directly by our clients and students. Sometimes the possibility of such pain is even vigorously denied. In addition, inferring social pain in clients and students can be difficult because some behaviors motivated by social pain are not those typically associated with pain (e.g., cutting off; intellectualizing; acting with hostility, coldness, prejudice, or indifference; interacting on the basis of perceptual distortions). Added to that, our professional assessment can be hampered by our own pained involvement and interactions with clients and students, individually and in groups.

I especially want to explain why I think our interventions with respect to social pain must move past our understandable inclinations toward sympathy and alleviation when in the presence of someone who is hurting. Yet, as a professional, it can be extremely difficult to maintain and facilitate an attitude of calm understanding, given pain's physiology, just as it is difficult for our clients and students to suffer this common human condition long enough for the pain to recede without engaging in habitually destructive patterns. Resolution can thus feel counterintuitive.

Aspects of social pain theory

"Social pain" is a term employed by social-cognitive researchers who have used functional magnetic resonance imaging (fMRI) of the brain to detect that the same regions activated during episodes of *physical* pain are also active during experiences in which research subjects reported feeling rejected, ostracized, or excluded (Eisenberger, Jarcho, Lieberman, & Naliboff, 2006; Eisenberger & Lieberman, 2005; Eisenberger, Lieberman, & Williams, 2003; Lieberman & Eisenberger, 2005). The term "social pain" is used to distinguish it from physical pain and especially to underscore its relational context. Additionally, researchers have found that the social pain response appears to be relieved by use of the right ventral prefrontal cortex, in other words, by thinking more objectively and less reactively (Eisenberger et al., 2003).

There are four interrelated aspects of social pain that I believe are important for understanding this phenomenon. The first is the automatic impulse from the brain's activated pain matrix to take rapid corrective action and then to learn from the experience. The second aspect concerns separation distress, particularly evident in infants when separated from primary caregivers but also noted in such colloquial phrases as "a painful breakup" and "a heartbreaking loss." The third is the physiological arousal that psychiatrist Herbert Thomas (1997) called "the shame response to rejection." And the fourth aspect involves the correlation between affective and interpersonal violence.

Pain matrix activation and nonconscious learning

The *pain matrix* refers to the brain structures typically associated with physical pain—the dorsal and rostral anterior cingulate cortex, somatosensory cortex, insula, periaqueductal gray, and right ventral prefrontal cortex (Lieberman & Eisenberger, 2005). When viewed using fMRI technology, these same structures also "light up" in response to rejection or even its perceived threat.

Although Berne did not write much about pain (and, given when he lived, would not have known the concept of social pain), he was still clearly aware of the power of this interpersonal dynamic when he wrote, "In my experience, a considerable number of hospital admissions have taken place shortly after the patient was told by a loved one (or even a hated one) to drop dead" (Berne, 1972, p. 111). Social pain can be that disabling.

To account for the intensity of such reactions, social pain researchers hypothesize that, in evolutionary terms, the preexisting physical pain matrix was co-opted for signaling social pain because group inclusion has historically been critical for human survival (Williams, Forgas, von Hippel, & Zadro, 2005). In the face of a life-threatening rejection, a person wants to act immediately. As with physical pain, social pain reactions ensure similarly rapid, automatic corrective action and also lead to similarly aversive learning for the purpose of avoiding future harmful social behaviors. This is analogous to quickly removing one's hand after touching something hot and taking care never to touch anything like it again. The avoidant behaviors are conditioned and likewise later triggered automatically and nonconsciously. Not surprisingly, this capacity for quick action typically leaves little room for thinking or dialogue.

Separation distress

Neuroscientist Paul MacLean (1990), who proposed the idea of the human *triune brain*, once commented that "a sense of separation is a condition that makes being a mammal so painful" (MacLean, 1993, p. 74). This is especially true if remaining in close contact with significant others means the difference between life and death. That urgency for contact and the corresponding threat associated with separation is expressed by mammalian infants in the form of acute cries of separation distress, cries that Panksepp (1998) linked to states of intense pain that are alleviated by the release of opioids in the brain. Moreover, such distress and pain associated with separation and loss are clearly active, to varying degrees, throughout the life span, as are sensitivities to rejection. Yet the fact that social pain is caused not just by rejection, but also by separation, can be confusing. Rejection, after all, carries a more hostile connotation, whereas separation can be seen more neutrally as mere physical distance, differing, or even individuating. What links rejection and separation to social pain is not so much the intent of the person acting in a rejecting or separating manner but the degree to which the person in pain experiences a threatening interpersonal rupture.

This may explain why the intensity of social pain appears directly correlated with the significance of the person or persons doing (or even witnessing) the rejecting (Thomas, 1997, p. 16). Thomas also noted that the pain intensity appears inversely correlated with the degree of psychological maturity, that is, the capacity for psychological "separation from significant others, particularly those of childhood" (p. 24). This maturational challenge has also been observed by researchers studying chronic physical pain (Eisenberger et al., 2006; MacDonald, Kingsbury, & Shaw, 2005). Thus, working with social pain dynamics requires a developmental perspective in which learning to manage and reduce such pain requires some maturing on the part of clients, students, and professionals alike.

The physiological shame response

Activation of the pain matrix, especially in socially agonizing situations, is also associated with pronounced physiological arousal throughout the body. In fact, Darwin (1872), in his writing on the emotions, prefigured the link made by contemporary researchers between social pain and reported experiences of shame, humiliation, and embarrassment as well as between social pain and observable physiological responses such as blushing, elevated heart rate, collapsed body posture, averted gaze, and blunted cognitive capabilities. As English (1975a) noted in her classic paper "Shame and Social Control," "To feel shame is to have a psychosomatic reaction" (p. 24).

Ernst (1971), writing about the *life positions*, also suggested a complex relationship among shame responses, feeling rejected (or inferior), and flashes of anger. He described "'Blushing,' 'Embarrassed,' 'Feel Foolish,' 'Self-conscious'" (p. 41) in connection with "I'm not-OK—You're OK." He listed "'A Burn' ... 'Red-in-the-face,' 'Hot-under-collar'" (p. 41) in connection with the more aggressive "I'm OK—You're not-OK" position. These apparently contrasting positions, in some cases, turn out to be physiologically and psychologically similar.

Although I am citing writers whose references to shame responses are consistent with social pain reactions, I do not consider shame and social pain to be equivalent. The physiological shame response is just one aspect of social pain. And shame is a culturally and psychologically complex topic in its own right, about which volumes have been written. For anyone interested in a concise overview of the literature, Erskine's (1994) is still useful today. A full issue of the *Transactional Analysis Journal* (O'Reilly-Knapp, 1994) was also devoted to the theme of shame, partly in response to a 1993 ITAA conference that addressed the topic, papers from which were collected and edited by James (1993). Yet some of what has been written about shame can illuminate the phenomenon of social pain, with which I am concerned here. In this chapter, therefore, I simply want to note, as many have observed, that shame responses are often a sign of an interpersonal rejection. And as I will elaborate in a moment, the violence of rejection itself can be paradoxically an outcome of shame and social pain.

Affective and interpersonal violence

As social pain courses through the body, there appears to be a direct correlation between the violence with which it can be felt and the violence of the action taken to eliminate it. Hyams (1994) wrote, "The deeper the shame, the more violent the hatred" (p. 263). Darwin (1872) also observed in humans and animals this "tendency to violent action under extreme suffering" (p. 72). He added, "Great pain urges all animals, and has urged them during endless generations, to make the most violent and diversified efforts to escape from the cause of suffering" (p. 72).

Escaping from affective intensity and suffering—especially the pain arising from interpersonal rejection, humiliation, or unbearable loss—is also what Gilligan (1996/2000) understood as underlying many of the most extreme acts of interpersonal violence. He came to his conclusions, like Thomas, after years of working clinically with prison populations in the United States.

Yet not all pain-activated aggression and violence is outwardly directed. Sometimes it is turned against the self, for example, when someone cuts his or her body to reduce unbearable social pain or engages in desperately corrosive self-recrimination in response to being ostracized. At other times, the violence takes the form of an extreme withdrawal or avoidance. Thomas (1997) wrote, "To avoid this pain, the person ... may withdraw and live a solitary existence ... to such a degree that in time others see the person as eccentric, the person becomes paranoid, or ... is diagnosed psychotic" (p. 24).

As might be expected, all of these behavioral responses appear to be accompanied by cognitive collapse or impairment, making it extremely difficult to access the Adult ego state, although, as I will discuss later, Parent and Child ego states can play a supporting role in what is primarily a nonegoic process.

Toward a more transactional model

Social pain research has focused primarily on the individual's physiology, response, and behavioral repertoire. The function of social pain is likewise largely conceptualized in terms of its survival benefit for the individual. However, to understand social pain as an interpersonal process, I believe the group's role must be considered, especially given the group's composition of individuals with their own activatable pain matrices. I am using the term *group* here, as elsewhere in this book, to mean any socially significant collective, whether a family, organization, or community.

The social function of rejection

Rejecting another person and causing pain seems like something we would want to avoid. Yet if individuals have evolved to avoid being excluded from the group, then the group's rejecting impulse must have coevolved for at least partly adaptive purposes. It would seem that groups benefit from remaining grouped, although not unconditionally. That is, just as an individual organism must sometimes eject or extrude what is potentially dangerous (infectious microbes, bad food, maddening thoughts), so too must groups sometimes conclude that their viability depends on getting rid of problematic individuals. Berne (1963) observed that "if [a] threat is too great, the group will attempt to extrude the agitator" (p. 20). He elaborated, "Extrusion across [the boundary, by the group,] is called expulsion, excommunication, or discharge; and exclusion is known by such terms as failure or rejection" (p. 57). Of course, the ostensible benefit of such extrusion or rejection will depend, in part, on the accuracy of the group's threat assessment, which is always questionable given pain's "act first, think later" strategy.

Yet not all group problems or threats require such extreme measures. Sometimes groups simply need to close ranks and quickly bring errant members back into line rather than ejecting them. That can be achieved with the nonconscious aversive conditioning or learning that the social pain response so efficiently produces. Pain-cued learning can thus sustain the group's coordinated efforts and activities. Other species have developed similar automatic rules and behavioral systems to reinforce and coordinate swarming, shoaling, flocking, and herding. English (1994), for example, suggested that "sham[ing] represents a particular evolutionary development which has an important role in maintaining standards in human societies" (p. 110).

This norm-shaping role of social pain and certain shame experiences highlights an additional cluster of pain-motivated behaviors: those involved in conforming and socially controlling. For although the literature on social pain and shame is full of stories of avoidance and aggression, little has been written about the aversive conditioning that can occur after a pain response. Conforming and socially controlling behaviors can be, of course, far less dramatic than avoidant and aggressive behaviors. One might not even associate them with social pain because the individual appears to be included and the group appears to form a united front. But although everyone may be on their best behavior, apparently harmonious relations may be predicated on conforming under threat of severe pain for deviating or individuating. In such cases, we see little flexibility within the system, either behaviorally or cognitively. And as with the more dramatic behaviors, individuals and groups caught in a pain-motivated mode of conforming/controlling find it almost impossible to reflect or make different choices.

Reciprocally activated pain

Although we have evidence that individuals experience social pain when rejected by significant others, what happens in the pain matrices of those doing the rejecting? If we recognize that everyone in the group has some sensitivity to social pain and that individuals often react to social pain by taking avoidant or aggressive action, we might reasonably infer that the amplification of those dynamics at the group level could lead to shunning, enraged hostilities, or other extreme forms of rejection or control. So, although to my knowledge no fMRI studies have been done at a group level, I hypothesize that the group's rejecting or controlling behaviors are themselves the result of social pain.

Consider the classic situation in which a group rejects (or attempts to control through threat of rejection) an individual perceived to be a deviant threat. That individual might understandably feel pain and react accordingly. But if we look at the situation just moments before the group's rejecting response, it is conceivable that the group felt intensely pained and threatened by the individual's deviation or separation from the group. Lewin (2000) made a similar link between the individual's separation distress and the violence of then paradoxically shunning significant others. At a group level, we might think of the group's nonconscious "violent and diversified efforts" as attempts to reduce and correct that painful threat of separation. If the perceived threat (and the attendant pain) is too great, the individual will be extruded. Otherwise, the rejection will produce not just pain but corrective action and aversive learning. And if all goes well, according to the logic of social pain, the system's equilibrium will be restored. Not only will the individual learn not to do that again, but the group will also learn, and its collective pain matrix will be activated when cued by certain preconceived behaviors, whether or not they actually constitute a threat. As we all know, the consequences of a misperceived threat at the social and political level can be tragic.

Yet once social pain is activated in this circular, systemic, nonconscious manner for everyone in a family, organization, or community, it becomes nearly impossible at times to identify an initial cause or culprit. Everyone becomes implicated in one way or another.

Professional susceptibility

To the extent that group member behavior is physiologically cross-regulated by social pain, these dynamics must also regulate dyadic behavior, as they surely do in the unavoidable intimacy of the working therapeutic or professional couple (Cornell & Landaiche, 2006). In Chapter 2 (see also Landaiche, 2007), I wrote of the *shared bodymind* as a fundamental human process of reciprocal affective and cognitive regulation—an interdependent, communicative, bodily process. This largely nonconscious interpersonal exchange occurs especially when learning and maturing is to be facilitated. It is a process, however, that can leave us, in our professional roles, highly vulnerable to social pain (Landaiche, 2014).

When we are not emotionally entangled within and activated by these rejection/pain dynamics, they can be relatively easy to identify in our clients and students. But once we, too, become caught and activated, we can participate in similar patterns of withdrawing, aggressing, or conforming/controlling. Our own "mental powers [can become] confused," to use Darwin's (1872, p. 323) phrase. We lose our professional neutrality.

In my experience, and in discussions with other professionals, our susceptibility appears related to three possible factors. The first is our inherited or learned sensitivities to rejection, that is, our alertness to possible signs of rejection and the intensity with which we then register social pain. That sensitivity can be activated by clients' and students' pain-motivated behaviors—whether withdrawing from us, acting with hostility, or comporting themselves so well or insisting on controlling us so tightly that nothing genuine can be transacted. Traditionally, such professional entanglements would be seen in terms of countertransference reactions.

The second factor is related to any actual dependence we may have on our clients or students for income, sense of job satisfaction, and/or referrals—in other words, for ongoing viability in our professional roles. Such dependence parallels that of humans in primary groups or families, in which inclusion and acceptance is crucial for survival. Any threat to our professional lives can likewise provoke an intensely pained response, along with the usual aversive and avoidant reactions.

The final factor is essentially an occupational hazard. Our work requires that we gather not only conscious, cognitive data (about the lives of our clients and students) but also nonconscious, highly emotional information (about their interior lives and affective relational fields). Due to the nature of this information, we have no choice but to gather it through somatic sensing or empathic resonance prior to processing it cognitively. As we do so, however, the other person's pain can become our own. Empathy, in fact, activates the same pain matrix that signals physical and social pain (Jackson, Meltzoff, & Decety, 2005; Singer et al., 2004). Once the pain matrix has been triggered—even if we have no reason to believe there is any actual threat of rejection—our automatic response to that pain kicks in, and we may find ourselves behaving in ways that impair our working relationships with clients, students, or groups.

Analyzing social pain transactions

Having laid out the complexities of this phenomenon as I understand them, I will now propose a simpler way of thinking about or analyzing social pain transactions.

First, we are not tracking the communication of cognitive contents or constructions from the ego states. Rather, we are looking at (1) a communicated or relayed rudimentary, visceral intention and (2) its reception and subsequent interpretation. That relay and reception/interpretation, however, is not necessarily symbolized or conscious. It occurs at the somatic or felt level. I think in terms of there being a "skin layer" for the ego states, encapsulating them within bodies that convey and receive messages, often nonconsciously, regardless of any particular overt interaction or content.

Relaying the intention

Viewed more schematically, there is an actor who relays an intention to separate from the recipient. I have chosen the phrase "intention to separate" to refer to a wide spectrum of possible actions, from pointedly rejecting the recipient, to conditionally loving the recipient, all the way to simply individuating from the recipient. Yet in using the single phrase, I want to convey that the spectrum and possible complexity of the actor's intention boils down to the same implication for the recipient experiencing social pain.

Moreover, the actor can relay the intention regardless of the overt content being communicated in words or behaviors from any given ego state. For example, I may be saying something that I mean to be positive, but my posture, the pace of my breathing, or the tone of my voice may be communicating anything from discomfort, to alarm, all the way to revulsion. Conversely, I may be saying something overtly critical or belittling that I am canceling out with emotional warmth, physical ease, or brightness in my eyes. Yet these differ from the understanding of ulterior transactions as occurring between specific ego states because the nonverbal cues just described originate and are conveyed bodily, in contrast to arising within and being transmitted from a structured ego state, conscious or not.

Receiving/interpreting the intention

Whatever the actor's more complex intention, the recipient's social pain response indicates that the separation was experienced as a threatening rupture or rejection. In addition, reception and interpretation occur simultaneously. That is, if the intention feels like a painful rejection, the recipient's interpretation is always that the actor intended it as such. Again, this interpretation is not made necessarily within a structured ego state. It is a primitive, instinctive conclusion made in response to a perceived threat. (See Ligabue's 2007 discussion about distinguishing among bodily processes, ego states, and script development.)

Bracketing other ego state content

This somatic relay and reception—this "skin layer"—functions as a barrier to the relay and reception of other communicative content from one ego state to another. This is because, in my view, any languaged or script-based content is secondary to the fact of the relayed and received/interpreted intention to separate. The somatic activation essentially overrides (and brackets) any ego state messages that are not consistent with the actor's relay of a separation and the recipient's pained interpretation of it. Although any number of other transactions could be co-occurring, I would not focus there first. Without resolving the social pain reaction, that more complex content cannot be usefully accessed nor can new decisions be made.

Role of the Adult ego state

In spite of the fact that the analysis of social pain transactions does not involve script analysis, per se, the Adult ego state plays a key role in resolving pain-habituated behaviors. But that role is not primarily one of evaluating and recoding script content (even if script content is implicated in the process). Rather, the main function of the Adult is first to recognize that pain is occurring (or has occurred) and that it likely accounts for a particular intransigent behavioral response. Yet the Adult's effectiveness in this case arises not so much in terms of *what* it thinks but more in terms of *that* it is thinking. To the extent that the Adult ego state represents activation of the cerebral cortex, such cortical activation apparently downregulates the activations of the pain matrix. It is only then —with the cessation of usually nonconscious pain and the recovery of more complex cognitive capacities—that one can begin working through the contents in other ego states and making new life decisions. Moreover, because social pain is collectively activated and regulated, one person's ability to remain in her or his Adult can stop the vicious cycle of pained retributions, helping to downregulate others' activated states as well.

Analyzing the transaction

When analyzing the social pain transaction, we place the client or student in the recipient position. The actor's role will be filled by a significant other, perhaps a parental figure or someone functioning as a therapist, consultant, or teacher. These static role assignments do not obviate the circular nature of social pain dynamics. After all, even recipients of rejection can engage in rejecting behaviors. But we have to stop the action somewhere. This simplification allows us to consider, in collaboration with a client or student, a single precipitating event in order to understand certain behaviors as reactions to social pain.

Once the pain reaction has been identified, we then figure out what the actor did to produce the sense of rupture. In my experience, unless a client or student is severely psychotic, he or she is often accurately aware of the actor's behavior. It is the interpretation that can be widely (sometimes wildly) off the mark. By first establishing the actor's literal behavior, the client or student can more easily recognize and understand the experience of being rejected and why certain reactions and behaviors may have followed from that experience.

From there it becomes possible to de-link the action from the intention (previously fused for the recipient). This means considering and imagining the actor's actual intention. Once thinking is restored, this shift of perspective is usually not difficult for the recipient because each of us typically has some insight into the person(s) whose rejection we have suffered. Moreover, as I will discuss in a moment, this shift to a more objective stance normally changes the rejection experience dramatically, with pain reduced to manageable levels, even in cases when outright rejection was clearly the actor's intention.

Challenges in practice

When social pain reactions underlie intense, intransigent behaviors, several challenges arise in attempting to change those behaviors: (1) identifying the signs of a social pain response, (2) embodying acceptance of the pain, (3) shifting to objectivity about the transaction, and (4) gathering courage through action in the world.

Identifying the signs

In most cases, we cannot rely on direct reports of social pain because, as noted earlier, clients and students are often not conscious of the pain and may be actively avoiding that awareness. Such avoidance may be literally, and understandably, an effort to get away from the pain. As such, social pain must often be inferred from a mix of physiological, behavioral, and narrative signs.

For example, in terms of the physiological arousal and nonverbal behaviors associated with shame, English (1994) offered this:

> Clues that the client is hiding a current issue related to shame even from herself, or, at least from the therapist, lie in fleeting changes in the client's facial and bodily expressions during a session, particularly if these differ from the person's usual demeanor or are exaggerated. Such clues include blushing, avoiding eye contact, hanging the head, wriggling in the chair, slight stuttering or lisps, halting speech, and inappropriate giggling.
>
> (p. 117)

These physiological and behavioral clues can emerge as the client or student talks about her or his life or interacts with the professional.

In my experience, the most significant indicators of social pain are found in clients' or students' narratives, which may include descriptions of interpersonal avoiding, blaming, attributing hostility and rejecting intentions, behavioral rigidity, and intense efforts to conform for or to control others (including the professional). Although certain aggressive and violent actions are often not associated with pain, they may be enacted precisely to avoid and triumph over the social pain, which would include some sociopathic behaviors.

Even in the absence of outward signs, our professional use of empathy can allow us to sense pain in a way that can feel highly unpleasant. And we may sometimes find ourselves caught in a social pain transaction with a client or student, one that triggers our own automatic, aversive behavioral sequences. I am often initially oblivious to such pained reactions, so I now monitor myself for certain typical behaviors, body language, physiological signs, and emotional states. For example, when I feel hurt, I become distant and rigid. I lose my capacity to think objectively about my client or student. I may start fantasizing about or actually working toward premature termination. I can feel faint, hot in the head, nauseous, chilled to the bone, frightened, ashamed, paralyzed, or inclined to lash out in harsh, critical ways. I become obsessed with thoughts of failure and professional ruin.

Yet in addition to managing my reactions, I also consider what they might tell me about the lived experiences of my students and clients and what I might anticipate in terms of our ongoing interactions.

Embodying acceptance

As professionals, when we see someone in pain, our first impulse may be to offer relief. We may instinctively move to soothe. Sitting with someone whose pain is not easily relieved can be its own kind of agony. Think of being unable to calm a colicky baby. "It is easier," Evans admitted, "to get active, to problem solve in helping the client to feel OK, than it is to actually live" (Erskine et al., 1994, p. 83). An impulse to soothe, however, may not best serve our clients and students in the long run. Relief may not help them manage their lives the next time pain comes along. As Mellor (2007) suggested:

> We remain helpful at the deeper levels with our clients only while we continue to face, accept, experience, deal with, and digest what is stimulated in us as we engage with our clients and do all this in a way that communicates our complete acceptance.
>
> (p. 176)

The challenge of embodying and modeling acceptance, therefore, lies in managing our own pain by not ejecting it (and thereby rejecting) in turn. In a previous chapter, I wrote about this process in terms of what Bion called *containing*—the capacity to remain in the presence of disturbing bodily experiences long enough for their elements to be mentally organized. Today I would trace most of my failures as a teacher, consultant, and therapist to my intolerance of sitting with clients or students in pain and especially with their pain-induced rejections of my efforts to help.

Shifting to objectivity

I think of deep acceptance or containing as a prerequisite for taking an objective stance or attitude. I am not talking about omniscience, however, but a willingness to look at social pain in its larger context. Thomas (1997) considered this objective perspective to be crucial for the client or student working to resolve a painful shame reaction. English (1975a) wrote that "often simply identifying the feelings and behaviors we feel most ashamed about and confronting them with Adult examination—our own and that of others—can reduce much of the inner agony and incapacity to develop self-confidence/autonomy and, ultimately, intimacy" (p. 26). Hyams talked of this in terms of trying "to take a little distance, to individuate ourselves from the problem" (Erskine et al., 1994, p. 83).

Most importantly, Thomas (1997) considered objectivity to include imagining the possible psychological state or inner life of the rejector. Goldberg addressed this when he asked, "What did your teacher [who shamed you in first grade] feel? Maybe she felt shamed by you" (Erskine et al., 1994, p. 81). In Thomas's experience, that effort to imagine the other's world resulted in the even further lessening of social pain, which typically occurs when analyzing the social pain transaction, as described earlier. Additionally, given that such transactions are often complex, the movement away from blaming (or from absolving responsibility) helps the client or student to be more accepting and objective about his or her own rejecting behaviors.

In my experience, for example, individuals who work to understand themselves within their larger family systems can, in time and with increased thoughtfulness, reverse habitual pain-motivated behaviors, such as blaming, attacking, withdrawing, cutting off, loving conditionally, and conforming for the sake of being loved. That shift from social pain to objectivity is essentially the movement from violence to insight.

Gathering courage

Talking and intellectual insight are not always sufficient to help our clients and students make changes. Sometimes they have to learn from living in the world. English wrote about a similarly active process that she called the deshaming "antidote treatment" (1977, pp. 298–301; 1994, pp. 115–119). This paradoxical, almost counterintuitive procedure involves consciously entering into situations "outside the safety of therapy" (English, 1994, p. 118) that have historically felt most painful. Behaviorally, one might think of it as a form of systematic desensitization. But the outcome is not numbness or sensory habituation. On the contrary, by reentering situations with acceptance, objectivity, and courage, it is possible to overcome nearly paralyzing fear and to be more alive with others. English viewed the goal, over time, as devising these antidote exercises "so that the next occurrences are less painful and the next doses of antidote more effective and can even be administered by the client herself or himself" (p. 117).

The choice to use an antidote exercise or to take any action outside of the professional setting must *not* be a function of the professional's discomfort with the client's or student's pain. Courage is not a function of professional direction, challenge, or pressure. Nor is it a process of toughing things out. The choice must emerge from the individual's determination to reverse a pattern of withdrawal or rejecting. Courage is strength that comes from personal conviction—an innate resource, albeit rarely one as instinctive as our capacities for rejecting and social pain. Our courage must often be willed into life.

Closing remarks

Because social pain operates in conjunction with many other developmental, environmental, and inborn factors, I have not found that working with it professionally requires a different set of techniques, per se, but rather a different kind of awareness, self-management, and patience. Human beings have been aware of social pain for thousands of years, yet our understanding of the physiological processes is relatively new. I find that the scientific perspective offers a way of organizing disparate psychological and human behavioral elements. These insights into an age-old human experience help clarify, in turn, what our clients and students are struggling with and what might be of use to them in their efforts to grow. Theoretical frameworks—of the kind provided by transactional analysis, object relations, and family systems theory—offer a way of usefully calming anxiety and, in effect, downregulating the painfulness of certain human interactions. I believe we are then freer to move forward with what Berne (1972, p. 128) conceived as the arrow of our sometimes-daunting aspirations.

5
LOOKING FOR TROUBLE IN PROFESSIONAL DEVELOPMENT GROUPS

To say that someone is "looking for trouble" implies, colloquially, that he or she is to blame for and deserving of any trouble found. Yet the phrase can also describe, more neutrally, searching for a particular trouble or problem in order to discover and learn from it. When I speak of my own "looking for trouble in groups," I mean it in both these reproving and researching ways.

A perverse pursuit

I never planned to teach. That role was not in my script (whereas becoming a therapist was). Yet when I reached my early fifties, I felt keenly how little time I had left to live. My familiar life felt somehow over. As if biologically triggered, interest in myself and my own growth shifted to a concern for the growth and achievements of those who would live beyond me, an urgency to help ready the next generation that seemed as strong as the impulse to preserve my own life.

Accompanying this urgency—to father, mentor, and teach—was a noticeable decline in memory and conscious cognitive abilities. My mind no longer functioned as before. I felt alarmed and frustrated, unable to organize my experiences and learning as I once could. Yet even without planned, careful thought and speech, something coherent was happening within me and with my clients and students. I just could not explain it.

With this mix of urges, racing time, and complicating cognitions, I found myself teaching in groups in a manner that routinely upset me. I pleaded, "Why do this to yourself?" For unlike being a therapist, which I loved and did to earn my living, teaching earned me little, and I did not love it. In fact, I frequently disliked my training role intensely. Yet I would not work in ways that might be more enjoyable.

In ego state terms, my Child complained, "No fun, too scary," as my Parent chided, "How could you?" with no real interest in the answer. Those became powerful motivations to jump that particular ship. But another, cooler part thought there was method to be found in my apparent madness. Its voice prevailed, much to my annoyance and much to my relief.

Mercifully, I did not have to make this sense wholly on my own.

Guidance for learning to teach

Where I live and work, teaching is something thrown to the elders, mostly by dint of practice experience and typically with little by way of preparation for the unique activity and process that is teaching. So I have been particularly grateful to the tradition in transactional analysis that takes the process of training as seriously as any other mode of practice. As Tudor (2009) observed, "Few other approaches ... have the same kind of career development and [training] structure that transactional analysis and its international accrediting organizations offer" (p. 276).

Education was featured in some of the earliest issues of the *Transactional Analysis Journal* (Ernst, 1971; James, 1979; Schlesinger, 1978), where it was already becoming sequestered under the catchall designation of "special fields" (see Clarke, 1981; Cornell, 2013). Over 30 years later, a full *Journal* issue (Newton, 2004) was dedicated to education and included a detailed literature review and developmental history of the field and its struggle for differentiation within transactional analysis (Emmerton & Newton, 2004). This was followed five years later by another full issue dedicated to transactional analysis training (Newton & Napper, 2009).

Consideration of the educational metaprocess—separate from a focus on technique or content—has emerged more gradually. Crespelle (1988), inspired by the work of Fanita English and drawing on Berne's (1968b) "staff-patient staff conference" model, articulated an approach to training psychotherapists that involved active observation and reflection on process. Newton (2003, 2014) examined theories and philosophies of adult education from a transactional analysis cultural script perspective. Grant (2004) looked at the way the transactional analysis approach to training correlated with the principles of effective adult education, ideas that Tudor (2009) explored further by considering the way "teaching 'in the manner of' TA" (p. 276) distinguished transactional analysis training activities from more standard educational approaches. By way of example, Pratt and Mbaligontsi (2014) wrote about using transactional analysis concepts in psychosocial support programs for grassroots community care workers in South Africa. Lerkkanen and Temple (2004) were interested in the effects of transactional analysis content and processes on the dynamic and interactive development of teachers personally and professionally. Similarly, van Poelje (2004) analyzed the experiences that led to developing successful organizational leaders in her effort to articulate a set of optimal learning opportunities.

Within the past ten years, Newton (2011a) has reflected on the nature, necessity, and ethics of risk in educational settings and processes, including the "risk of challenging the system" (p. 114). She (Newton, 2012) also wrote about the integration of learning models for the supervisory process, as did Chinnock (2011), whose approach to learning in supervision is strongly informed by the emerging relational approaches to transactional analysis (Fowlie & Sills, 2011; Hargaden & Sills, 2002b). In the spirit of acknowledging the commonalties among different fields of practice, Joseph (2012) argued that Berne's (1966) "therapeutic operations" (pp. 233–258) could also be conceptualized as interventions relevant to teaching, akin to Newton's (2011a) conception of "real learning as a therapeutic process" (p. 114). Such transformational depth was also elaborated by Barrow (2011), who drew from personal learning experiences to elucidate the intimate connection between education and our human gifts for cultivating what we need for living and growth. He later linked these themes to the somatic and environmental impacts of educational encounters (Barrow, 2018).

Some significant aspects of this community endeavor have since been gathered in several books devoted to educational transactional analysis (Barrow & Newton, 2016) and to training (Barrow & Newton, 2004/2013; Hay, 2012; Napper & Newton, 2000/2014). As Barrow and Newton (2016) wrote in the introduction to their edited volume:

> Education involves an act of faith. When we gather in the education endeavour, teachers and students step into unknown territory; a creative process that has potential to change, renew and transform. Ultimately, education is a way through which humankind expresses an enduring capacity to thrive. In turn, educational transactional analysis provides a psychological framework for explaining what happens in learning, how teacher and student roles interact, and illuminates why educational processes are, as Yousafzai claims, "the only solution" to how individuals act to change the world.
>
> (p. 1)

I was so affected by reading this book—with contributions by 22 authors from 14 countries on 6 continents—that I wrote a review in which I described how

> these numerous points of sharp and stimulating contact remain embedded in me ... less as a set of ideas and more as a world of sensate others, each working in her or his way on a common project and passion. Those multiple points of stimulation and reference remind me that I am not alone in attempting to understand. Others are also striving to put into practice the methods and the

ways of being and interacting that will facilitate learning in those whose paths in learning cross ours.

(Landaiche, 2016b, p. 249)

In so many ways, these efforts and this tradition within transactional analysis have helped greatly in my struggle to articulate what I am doing when I am striving to facilitate learning in many contexts.

No single field

Although this chapter fits nominally into the field of education and training, I think Tudor (2009) aptly pointed out that "all transactional analysis trainers are, in effect, in the educational field of application" (p. 276), no matter their area of primary practice. Moreover, I think the process of training and education can elucidate the process of change that is fundamental to every field, an interdisciplinary attitude faithful to the origins of transactional analysis (Landaiche, 2010).

A succinct example of this interdisciplinarity can be found in Berne's (1968b) relatively brief paper on "Staff-Patient Staff Conferences." I believe it exemplifies the complexity of his overall project. For example, to manage the dynamics of that particular institutional group encounter, he applied a deep understanding of human emotional and psychological difficulties and demonstrated skill in attending to the here and now. He showed awareness of the formative effects of simply observing along with an acute attention to organizational and group realities and processes. Moreover, he structured the conference meeting with keen attention to the learning of everyone involved, whatever their official roles, along with an intuitive understanding of the systemic interplay of these varied human aspects—psychological, sociological, aspirational, and logistical.

In his body of work, Berne aimed for nothing less than a comprehensive understanding of humans as social beings and of the elements necessary to guide our productive, satisfying growth. Since Berne's death just 50 years ago, the transactional analysis community has continued pursuing that aim. Due in part to that collective venture, I can now spell out my assumptions for developing professional capacity, particularly in what appears to be the unavoidability of groups.

Impediments to growth

As I wrote in my review of *Educational Transactional Analysis*, edited by Barrow and Newton (2016), "The [TA] educational model's emphasis on growth helps me keep in mind that the distress I encounter on a daily basis is, above all, a profound disruption in the process of learning, of integrating human experience" (Landaiche, 2016b, p. 250).

In my work as a teacher, therapist, supervisor, and consultant, I have become intently interested in signs of learning and growth because I now see learning as the primary activity of being human, an ongoing, sometimes harrowing process. So in my different professional roles, whether working with individuals or groups:

> I am called to attend, moment by moment, to the effects of history and social context, to the effects that my way of being have on the sometimes tiny movements of becoming. I am motivated thus to modify or stabilize, to discover what it is that my client or trainee is seeking to learn—teaching me, in effect, how to teach, how best to facilitate.
> (Landaiche, 2016b, p. 250)

I am also interested in what makes being a human so difficult, so disruptive to growth and potential. In previous chapters I wrote about what I see as the sometimes considerable trouble of having a human mind with its human body and about the difficulties that this condition presents for the work we do as human relations professionals. Moreover, as I proposed, I see the troubles that bring people for help and that make helping difficult can usefully be traced to two basic, interrelated human challenges: (1) our capacity for thinking what is not necessarily real and (2) our incapacity for bodily affect and responsiveness to living.

Life requires each of us to organize a wealth of experiential information, absorbed into our minds and bodies, a flood of data that, when incorporated, gives us the ability to take directed action toward our dearest aspirations (Berne, 1972, pp. 128–131; Clarkson, 1992, pp. 12–13). Yet without a capacity for neurophysiological integration, we cannot grow in that forward direction.

This seems certainly true, for example, in situations wracked by the social pain dynamics I described in Chapter 4. Consider school settings where there is bullying or shaming—whether by fellow students or by teachers. The entire school atmosphere may be saturated to the point that few individuals may be able to acknowledge much less remediate what is happening. This will have profoundly negative consequences for the learning of everyone implicated in these systemic dynamics, not just for the targeted individuals.

Thus, when excessive affect or stimulation makes it hard to consolidate experiences, the gaps between reality and our cognitive schemas widen, and our decision-making becomes increasingly impaired. Berne (1947) referred to these tensions of the human energy system, prosaically, as "the problem of a human being" (p. 42).

Such human problems can be the traumatic result of both catastrophe and aspirational growth—in short, the consequence of being alive in our responsive, dynamically reorganizing, largely nonconscious bodies. In

Kolb's (1984) words, "The challenge of lifelong learning is above all a challenge of integrative development" (p. 209). I believe this holds true for groups and communities as well.

And so the student unable to assimilate the emotional implications of new learning will not be helped by a teacher unable to consider emotional factors involved in making new meaning out of information. The client unable to live with a distressing experience may likewise encounter a professional unable to tolerate such force and anguish. A client organization unable to adapt to the reality of changing economic circumstances that provoke profound emotional reactions within that organization cannot be helped by a consultant also unable to sit with that level of threat, tension, and overwhelming data long enough to think within it. As organizational consultants, Lawrence and Armstrong (1998) wrote, "Non-psychotic thinking cannot be entertained until and unless the experience of being pulled toward psychotic thinking is experienced, internalized, and worked through" (p. 55). That is, until we can work within students' or clients' emotional, cognitive turmoil, we cannot begin to understand what it will entail for them to suffer and master such disturbance in their lives and work.

I believe these difficulties also mark our developmental impasses as professionals.

Developing capacity

When I speak of development for human relations professionals—consultants, trainers, and clinicians—I mean quite specifically the development and growth of a particular capacity, relevant in dyadic contexts as well as in work with multiple individuals.

When we hear the word *capacity*, we may think of abilities or skills, which may differ in each field or discipline of practice—that is, the capacities to do the job we have each been trained to do. But I am drawing here on *Webster's* (1975) definition of *capacity* as "the ability to hold, receive, store, or accommodate ... the measured ability to contain" (p. 164). I want to use the word *capacity* in this narrower sense to describe a competency common across fields. I want to convey a bodily attribute, as in a capacity to contain or hold, an ampleness, a specific ability to suffer human experience, to think in its midst, a large enough ambit of being in our own skin that makes room for what is still intolerable to those coming to us for help.

One might say that the outcome of a successful professional engagement with troubled students or clients is the development of their own capacity to bear experience, to persevere in thinking about it, and thus to be more amply present in their lives and ongoing learning.

So, although I am interested in many kinds of learning, and in the learning of many kinds of things, in this discussion I am specifically interested in the learning that develops professional capacity, more a process than a content, that is, a process ability that makes room for many potential contents: skills, stories, procedures, data, meaning, discoveries, and so on.

Emotional learning in unavoidable groups

Learning and education are often framed in primarily cognitive or intellectual terms. But I think it is important to talk about learning as being an often less-than-rational process. As Brooks (2009) described it, "The knowledge transmitted in an emotional education ... comes indirectly, seeping through the cracks of the windowpanes, from under the floorboards and through the vents" (p. 9). That is, if we attend to the extraordinary diversity of experiential and sensory data, there is more going on in our learning environments than fits the stated objectives.

When Newton (2003) wrote, "Transactional analysis began as a group therapy and to a certain extent it still is" (p. 321), she pointed to Berne's (1961b) characterization of transactional analysis as an "indigenous approach derived from the group situation itself" (p. 165). Newton was also referencing an aspect of the transactional analysis tradition in which significant emotional learning is seen as fundamental to the work in any group promoting learning, growth, and the recovery of what Berne (1964) called the capacities for "awareness, spontaneity and intimacy" (p. 178).

In many contexts and areas of application, group work has played a significant role in the transactional analysis tradition, for which I provided an overview in Chapter 3. And for the simple reason that a great deal of transactional analysis training occurs in groups—workshops, seminars, conferences, reading groups, supervision, peer consultation—they might appear simply unavoidable. Yet I do not think this ubiquity can be attributed just to economies of scale. Rather, I believe that certain learning about being a human among other humans can only occur in groups, the context that Bion (1959/1969) believed was quite as "essential to the fulfillment of a man's mental life ... as [the group] is to the more obvious activities of economics and war" (pp. 53–54).

This essential, unavoidable learning context seems closely linked to the problems groups present: limited time and resources, disorganization and chaos, conflict, banal groupthink, peer pressure, affect intensification, boredom, passivity, scapegoating, and what Bion (1959/1969) called *basic assumptions*. To my mind, the unavoidability of groups is part of what creates their tensions, mirroring similar struggles we may have in community, family, and organizations—the groups without which our personal and professional lives would seem immeasurably impoverished if perhaps simpler.

Such learning in groups and communities is rarely trouble free. Berne (1964) even ventured that pursuing meaningful growth might be "frightening and even perilous to the unprepared," adding that "perhaps they [would be] better off as they are, seeking their solutions in popular techniques of social action, such as 'togetherness'" (p. 184). He then concluded, "This may mean that there is no hope for the human race, but there is hope for individual members of it" (p. 184).

With that caution in mind, I approach a training group assuming that all present have a capacity to learn, areas of learning difficulty, and priorities for growth and development. Yet individuals may not choose to pursue these. I expect the group to facilitate the more manageable learning and to block, with anxious ferocity, the most difficult areas. My job is to attend to the learning aspirations and impasses of the individuals and of the group, all the while remaining in contact with my own such yearnings and deadlocks.

Contracting for learning outcomes

In training, as in professional practice generally, there is typically a contract and an effort to fulfill it. The administrative and professional aspects of a training contract, for example, involve place, time, fees, themes, activities, number and profile of participants, generalized goals, the professional's comportment, and so on (see Berne, 1966, pp. 15–16). These can be straightforwardly communicated, agreed upon, and realized. But, arguably, the most important part of the contract relates to the learning objectives or outcomes, and that is where things become murky. As Hinshelwood (1983) ruefully noted, "It is the glorious privilege of academics to know that they are on the track of knowing everything. It is the humble gloom of the practitioner to know that nearly everything remains uncertain and paradoxical" (p. 167).

For myself as a trainer and group facilitator, I know in an all-too-general way that professional development is our goal, which I define in terms of (1) increased understanding of human functioning, (2) increased capacity for sitting with the fullness of human experience, and (3) increased awareness of how psychophysiological maturation is facilitated in different contexts. But what does this look like for the individuals who have come to learn? How will they articulate their position in relation to these abstract, lofty goals? I know that when I am a member of a professional development group, I may have ideas about what I want to learn, but I rarely know what I will actually learn until I am reflecting on it afterward.

As Maquet (2012) noted in his discussion of contracting for therapy, the "psychological contract" (Berne, 1966, pp. 16–17) is often much harder and in some cases impossible to pin down. Indeed, often part of every problem we try to solve is that we do not necessarily understand the problem: its scope, origins, or resolutions. These may be aspects that, in a group, will only be discovered as the work unfolds.

As a facilitator, part of my professional contract is a commitment to support and watch over whatever learning emerges. I maintain a boundary between professional development and personal therapy while trusting that each individual will also decide where to draw that line. I am there to help with professional growth knowing that much of the growth in our work comes from personal explorations and knowledge. Indeed, my own capacity and structure—my integrity as a professional and person—is part of what I agree to use in my work, even though I will not know its mettle until tested.

Although group members may feel intensely ambivalent about their own "frightening and even perilous" learning process, as Berne (1964, p. 184) called it, I err on the side of believing that each person comes with a potentially unarticulated wish to grow in relation to particular challenges and aspirations, just as the group collectively embodies its own such set. I work on faith, as it were, to figure out what those challenges and aspirations might actually look like.

So, as our work progresses, we continually renegotiate what I call the *developmental contract* (see Lee's, 1997, notion of the *process contract*). And we do this within the agreed-on administrative and professional framework of training and in relation to "the raw desire of individuals and groups to accomplish, develop, and flourish" (Barrow, 2011, p. 311).

The orienting questions

To guide our work in a group that is developing the professional's capacity, I propose two orienting questions:

(1) What learning is important for each of us?
(2) How can we each make use of this group experience to achieve that?

These apparently basic questions are surprisingly hard for many of us to ask, much less answer. Our learning experiences may have been structured by what others wanted us to learn. We may not have learned how to pursue and develop our own learning potential. We are then likely not aware of what kind of learning has worked for us in the past or how to secure the help we need. We may not have discovered how to attend to the particular capacities we long to embody in our work and lives.

Yet even in the absence of answers, these research questions direct us toward the natural force of our aspirations, for which our learning is most relevant. In effect, our engagement with these orienting questions represents the primary process structure of the groups I am describing.

I use these questions especially as a guide when members cannot articulate their wishes or needs, which is common and natural. When I see a sign of growth or learning, for example, I use that to begin constructing hypothetical answers to the questions. I infer a possible direction of growth. When I voice my hypotheses or inferences, group members' disagreements are sometimes the clearest sign that, at a bodily level, there exists a strong urge and urgency toward a particular hope for life, even if it is only known in terms of what is not wanted.

Aspiration is what moves the group in a purposeful manner through the upheavals of our disorderly process.

On abandoning the lesson plan

I am someone who likes to have a plan, and I invariably create one before I facilitate a learning group, even though part of me thinks such planning is pointless. Once I discover what people are really interested in learning—within the frame of our agreed-on theme, for example—I have been known to scrap my agenda and pilot entirely by feel.

One can imagine the potential for chaos in groups of this kind, which a colleague ruefully referred to as Brownian motion, the random, frenetic movement of single cells or dust particles in the air. His statement in the group context was a moment of creative linking that integrated some of his disturbance within that very chaos.

In the absence of an agenda, I still must structure the experience. Sometimes the structure is my capacity, as a leader, to take in disorder and be willing to think about it. I have come to trust that by listening to the group, by receiving its complex, multimodal (verbal and nonverbal) communications, my nonconscious self will be affected, and I will respond with an ethos of care, with attention to the signs of growth and impasse. I watch for psychosomatic integration—for thinking in its full bodily sense—and I watch for growth's destruction, that is, its failure to thrive.

I believe strongly in experiential learning (Kolb, 1984), in which the experiences of talking and being in a group of individuals pursuing their own urgent learning goals is followed, ideally, by reflection on that experience—putting words to what has happened for individuals in the group, "the basic human ... process of ... generating hypotheses" (Newton, 2006, p. 186). Those words and hypotheses become the relevant content.

Sometimes, however, when I am overly anxious as a group leader, I try to reverse that sequence. I deliver the content first and imagine that the learners can then infer the experience, in effect bypassing the difficulties of actual experiencing. In such instances, the intellect may be developed rather than actual, bodily grounded knowledge. We then have ideas about living and working rather than development in actual capacity.

The integrating potential of free speech

My experience in professional development groups has been that everything happening there can potentially and experientially teach us about human behavior, including how humans struggle with and help one another develop. So when I facilitate, I leave the process format open: open to what is important to each of us in any given moment; open to what is emerging (thematically, emotionally, interactively); open to speaking; open to what we do not yet know. I make time for participants to describe what they are thinking and feeling; what they are aware of; any questions, observations, and associations (images, stories, funny or not so funny linkages, etc.); as well as identifications or disidentifications with what has been said by others. At the same time, while invited to speak, group members are free to guard their privacy, thus free to be silent as a means of securing reasonable safety.

In a therapy context, when a client is asked to say what is coming up—what is coming to mind, what feelings are present, what bodily sensations are in awareness—this inquiry communicates the therapist's (1) intention to follow the client's lead and need, (2) faith in the client's internal knowledge and wisdom, (3) acceptance of the client's being, and (4) expectation of the client's active role in her or his treatment.

In a learning group, such openness and invitation seems to move the members most efficiently toward the problems to be worked on, investing their work with the motivating energy of emotion and meaningfulness. Freely associated material uncovers surprising connections and clues to otherwise puzzling, sometimes imponderable, aspects of the group's life. It offers freedom to move past a linear, singular focus on a symptom and a foregone conclusion about its resolution.

Although freely associating in mind and speech can be enormously liberating for some, it can be anxiety provoking for others, from bearable to disintegrative. In those worst cases, I ask people to think about what they need to do in that moment to recover themselves. I attempt to put what I am understanding into words, although it I find that what feels most protective varies: Sometimes it is my speaking, while at other times it is my capacity to sit silently with restraint (see van Beekum, 2006, 2012).

In the life of a learning group, there appear to be moments when conversations among group members—transactions across the group's circle—are productive. Shaw (2002) spoke of this as

> conversing as organizing ... a process of communicative action [with] the intrinsic capacity to pattern itself. No single individual or group has control over the forms that emerge, yet between us we are continuously shaping and being shaped by those forms from within the flow of our responsive relating.
>
> (p. 11)

When group members approach internal clarity, they speak directly about themselves to one another. Their words are loaded with affect and history, the more in-depth exploration of which I consider outside the scope of a learning group. So I continually return to the task of describing the immediate experience, immediate images and thoughts, apparently random associations, memories, fantasies. This not only surfaces important themes, it also functions as a form of conversation. People feel heard when their speech is echoed in another group member's speech, even when the linkages are more intuitive than apparent.

Just as often, however, people talk in groups as a form of discharge and escape, without interest in their own emotional motivations or in what others have to say. I find such transactions to be destructive to the group's capacity to discover, to describe, to sit with.

So, a key part of my task as facilitator is to safeguard opportunities for people to talk about the experiences they are having, to allow that communication to inform the next suggested learning activities and experiences, and so to grow from there.

These are judgment calls on my part that sometimes succeed and sometimes fail. Such risks seem inevitable in a process this alive and unpredictable.

Listening with my body

As a facilitator of professional learning processes, I often have no idea what I am doing or where we are going in any given moment. Yet I see this as more than mere incompetence. I see having no idea as key to my role. As Tudor (2007) noted, "The facilitator, too, is 'living the uncertainty of discovery'" (p. 387).

Yes, of course there are the logistics, the time boundaries, the contracts, the structure of learning activities, and the content as applicable. All of these are reasonable, necessary, important, and relatively uninteresting to me. Which is to say, all of them—time, contracts, structure, activities, content—simply create the occasion for the problems that emerge from learning in groups. And it is to these problems or troubles that I most strongly direct my attention.

I see my job as scanning for trouble.

The group begins its broadcast sometimes even before I have officially begun by calling the time. This broadcast is continuous, whether the group is chitchatting, silent, hard at work (or avoiding it), intimate, defended, regressed, maturing, or just plain dull. Berne (1972, p. 322) made a distinction between the professional's Adult ego state listening to the content of what is said while one's Child–Professor listens to the *way* it is said. Indeed, there are many instruments of such nonverbal transmission: body postures, qualities of eye contact, breathing patterns, restlessness, calm, agitation, fear, boredom, withdrawal, excitement. (See also Berne's, 1966, p. 66, detailed list of physiological signs to watch for.)

We receive these many signals with our bodies, often outside awareness. More so, I would argue that "before we had verbal language, as a matter of survival we knew how to read the signs expressed in the non-linguistic patterns of group life" (Schmid & O'Hara, 2007, p. 102). This innate urgency to notice, receive, and interpret the signals from others thus operates for small children before they have words, as it would have for us as a species long before we had language. Listening with my body refers to this primal human ability and inclination.

Listening carries the double connotation of hearing and attending: "When I attend to you, I stretch my ears toward you" (Alvarez, 2005, p. 18n). Thus, in listening, we actively invite what we receive. And our bodies listen with more than just our ears; we extend our full selves toward the other to pick up more than sounds, more than words, sights, smells, or sensations. When I listen with my whole body to the patterns and flow of the group—listen to the tempo and rhythm, the temperature and tensions, even as I actively contribute to them—I experience myself sinking into the group's field, its raw physicality. I submerge into its efforts, productivity, and struggles, which is often more than my poor little mind can handle. So I must rely on my full body, not only on its sensing capabilities—its somatic resonance—but also on its ability to organize, often without my knowing what or how.

To illustrate what I mean by listening with my body, I offer the following excerpt from the life of one particular learning group (disguised to protect its participants).

I am facilitating a professional development group for consultants, educators, and clinicians that is exploring trauma as a psychological phenomenon and as a community process. We are well into the afternoon of the first day. Members have been speaking and asking me questions, which I am trying to answer. Their speech is self-disclosing and exploratory as well as intellectualized and evasive. The questions express both genuine interest and an avoidance of thinking. I have the sense that someone is being blamed. Me, I wonder? Yet everyone is looking at the person who is dominating the conversation, who is apparently in charge but, it feels to me, actually quite anxious, in charge of nothing, frantic. While I speak and listen to the overt exchanges, I am watching out for the dominating one, alert to possible attacks. I am increasingly uncomfortable, unsure what the group actually needs from me—with needs and demands all tangled up. I am listening to what people are discovering through speech, finding myself surprised and a bit in awe. Other words of theirs feel dead. My breathing is constricted. Something is being lived that we have not come close to identifying. Hundreds of small gestures are being passed around the group: alliances signaled, disdain expressed, calls for help. In the midst of a long ramble by one of the more intellectually defended members, she inserts a small phrase that does not fit. It seems innocuous: "my grandmother's faith." I am trying to follow everything being said and enacted, but this phrase is like a small stone in my shoe. I find tears coming to my eyes. This happens as I notice another group member's face flush with feeling. Something drops in my belly; information is flooding in peripherally. I have the sudden notion, more a bodily rearrangement than a thought, that we are living what earlier had been just theory, just ideas, that we are recovering the overwhelming emotions specific to these individuals' experiences of trauma, that we are close to the grief that has felt untouchable. Yet this is not a therapy group. We have come to learn about an aspect of human experience that seems ordinarily unapproachable although relevant in nearly every work context.

Amidst this process, as I sit with the phrase "my grandmother's faith," I have no interpretation other than to link it to the transgenerational process, to the function of spirituality for our families and communities, and to the ways we sometimes also misuse our gifts. Yet rather than focus on that possible content, I follow instead the emotional opening offered by the phrase. I find halting words to describe what I think is happening: how it relates to our theme; how trauma can sever our trust in one another, without which we cannot move forward, singly or collectively; how in trying to master our reactivity we can repeat the horror we wish never to revisit; how hard it can be to forgive ourselves. Others pick up the cue, giving voice in ways that exceed my own way with words—sharing their own memories of family and of family history; their own shame and regrets; their struggles in regard to faith and hope; the connections they now see between challenges presented by their clients and what they, too, have lived or avoided. We appreciate this momentary reprieve. I can feel the difference in my body between the actual relief of contact and the phantom relief of escape. Then my chest tightens—I do not know why—as I shake my head internally: "No, no, no," aware of the miles and hours still to go, the problems still ahead, like the foreshadowing that precedes a difficult stretch of weather.

From my submerged, disoriented position in a group, if patient and tolerant of what I receive, my body begins to form a felt sense of where the group aspires to go, even if the group's ambivalence sometimes shows up as hostility or impasse. Often before I can put it into words, I can feel the drift of what the group is creating, something new that exceeds the capabilities of a single person. From this vantage of contact, it becomes more possible for me to participate in the group's striving to articulate itself, its yearnings, its unique collective perspective—its genius, if you will.

When to speak up and what to speak about

In regard to the educator's most essential function, Barrow (2011) wrote: "Act swiftly on the instinct that something may be amiss. Do not wait and wonder" (p. 311). He captured well, I believe, the urgency with which we must, in our roles as facilitators, prioritize safety above all else. So, at the first sign of someone being harmed during a group session, I call a halt to the proceedings, either to say what I am observing or that I am concerned that someone is being or might be harmed. I try also to link what is happening to our theme. In learning groups, I am speaking of psychological and emotional harm more than of physical violence. Harm at an interpersonal and intrapsychic level is sometimes hard to detect. And I can find it equally hard to intervene in a manner that supports growth without shaming participants.

To that end, I sometimes have to make myself the target of the hostility—perhaps asking the group if there is anger with something I have done or said, perhaps inviting the aggressor to engage instead with me. Ideally, I will intervene at a group level so as not to single out the aggressor, who may be acting on behalf of the "innocent," silent ones. My interventions are based on the hypothesis that some of the members' most intense feelings relate to the leader's incapacity to contain the experience and facilitate growth. In the absence of that security, anxiety and hostility will be directed instead at other members, who are seen as safer targets. By drawing the fire toward myself, I signal a willingness to make contact with what is being disturbingly acted out in that moment, perhaps to find words to describe it, perhaps only to sit with the sometimes profound discomfort of disorganized affect and unknown history.

In other, nonurgent situations, I mostly speak when the data of experience have organized themselves sufficiently to describe either the growth or the dilemma in the group—not necessarily shared by all but in some sense affecting all. So, for example, a growth in trust can be described at the same time that some individuals are much less trustful than others. Fear can be noted even when there is evidence also of courage or risk-taking. There may be indications of pain amid incidents of pleasure or satisfaction; clarity with confusion; envy with gratitude; a series of combinations, variations, and contrasts. I speak what comes together in me as a result of attending bodily to the group's life and its struggle. My speech is merely one voice among many.

Solomon (2010), in her moving account of being a patient in one of Berne's groups, wrote:

> Dr. Berne was quiet in group, often listening to the conversation with closed eyes. The act of closing his eyes took him out of the interaction and allowed him to be an observer, thus increasing the

level of interaction between group members. It also enabled him to focus on the subtle changes in voice that come with changes of ego states and the emergence of games and script material.

(p. 184)

I have found no way to accelerate or rush this process of organizing experience, my own or that of the group members. It can only happen organically, within the limits of the group's and my own maturation.

Looking for and making trouble

When de Graaf and Levy (2016) wrote, "We feel the constant challenge to develop and maintain a learning climate in our training groups that is as safe as it can be and as disturbing as it should be" (p. 229), they described well one of the primary tensions I believe we face as facilitators of learning processes.

For as much as I prioritize safety, I also see the areas of trouble in groups as the areas of greatest potential for growth. They are the impasses to be resolved, often the areas any of us would least like to look at. So I think of it as my job to venture into those forbidden zones—venturing as a form of inquiry, exploration, lying in wait for the trouble rustling in the nearby brushwood. My strong sense of commitment to going there is in strong conflict with detesting that aspect of my job. Some people like roller-coasters and horror films. I abhor scaring myself near to metaphorical death.

No matter how much preparation I do beforehand—in terms of content or the structure of learning activities—if the development of professional capacity is part of the mix, part of me is prepared to drop the preparation in favor of the group's process and the potential it carries for learning. As time goes by, I do such depreparing increasingly. I launch myself recklessly into the process: the unspoken, the dynamic of this group now, the areas of difficulty.

When the group appears to proceed smoothly and I am the idealized leader, part of me is relieved. Yet another part—sensing that learning may be the last thing happening in such circumstances—refuses to play along and introduces elements of tension, of missed cues for attunement (or collusion). I let things stir up. I feel the group begin to unravel, and fear seizes my gut.

This process of inviting rather than soothing trouble is well described by van Beekum (2006, pp. 327–328) with regard to consulting with an organizational group during which his own strong urges, feelings, and fantasies played a key role in the group's gradual understanding of a crucial organizational issue that was operating underground, repeatedly acted out in unproductive ways.

I likewise often find the trouble I am looking for and hold myself responsible for the trouble I get into, whether I am happy about it or not.

And if we are fortunate as a group, we come through this disorder and alarm; we move from nonlearning to learning, which is a relief, yes, but also a wonder, an occasion for another kind of gratitude.

If I could be guaranteed this fine outcome every time I led a group into this kind of miserableness, I would proceed with confidence, braced by the surety of success. But learning does not always occur. Failure is always possible. And when it occurs, I feel a piercing shame (Landaiche, 2014). That unwanted option, above all, haunts my repeated descents into any group's region of trouble. I have found no way around that risk or distress in my role as a trainer in groups.

Recurrent disarray and hope

Rogers (1955, p. 267) wrote about launching himself into a therapeutic relationship with the faith that his capacities would "lead to a significant process of becoming" for the client. Yet he also acknowledged that if what developed instead was "a failure, a regression, a repudiation of [himself] and the relationship by the client," he would have the quite painful sense of losing himself or part of himself. He recognized this as a very possible and present risk in his work and an outcome he would experience very keenly.

In similar ways, I can still find it terrifying to facilitate learning groups. The search for the areas of trouble, so often explosive and mean; the plummet into the realm of disorientation; the sense of being overwhelmed by the group's many affects; the embarrassment of encountering my limits and frailties in front of a group—these appear to be enduring distresses. I cannot seem to calm them, even knowing that living through them has never been as bad as my fright.

I have learned to suffer terror better but not yet to befriend it.

Of course, my job, as I conceive and execute it, is not all bad, even for an overly sensitive guy like me. I love the moments of intimacy and consolidation that occur. I like the surprising creativity that people can achieve in groups. I am grateful for the group members' many contributions, often more insightful than my own. I am especially grateful for my own continued learning, which would not have occurred outside of a group.

Yet such lovely moments seem inevitably followed by further ugliness, an arrhythmic passage between what one might call finesse and dissolution, an orbit stately then unhinged.

With contracts in flux, the group's dynamics following their own rhythm and tone, the learning process itself one of integration followed by renewed disintegration, it is a hazard of leadership to be continually in the position of not knowing at precisely those moments that feel most fraught and in need of knowledge. I feel called at such times to recognize the limits not only of my own learning and knowledge but also of my ability to facilitate that for others. This is humbling. Part of my job is to face that I may not be the right person for some particular jobs.

Yet,

> the arrival of hope and some kind of understanding of the work is perhaps the first novelty to emerge ... the first experience of help and improvement, the emergence of a different way of understanding changes and how changes emerge.
>
> (Christensen, 2005, p. 103)

My own learning

I have always said I love learning, but in fact I hate it almost as much as I hate being part of groups. It would be more accurate to say that I love the mastery that comes after I have gone through the often-miserable process of learning it. And as I elaborated in Chapter 3, learning in groups has been especially difficult for me (see also Landaiche, 2012). I have sympathy for members who share that challenge as well as envy and admiration for those who learn in groups with more grace or apparent enjoyment and ease.

Yet I have been grateful for the guidance offered by radical educator Paulo Freire's (1970/2000) contention that "the teacher is no longer merely the-one-who-teaches, but one who is himself taught in dialogue with the students" (p. 67). For indeed, learning how I learn in groups has been essential for my learning to facilitate them. As a result, it is now easier to keep track of my own difficulties and peculiarities as separate, albeit related to, what I am trying to discover about any particular group's development.

"Training for me," writes P. K. Saru, "is an evolutionary process wherein both trainer and participant mutually evolve, learn, and integrate within and outside the self" (Saru, Cariapa, & Manacha with Napper, 2009, p. 326). Indeed, what I observe is that in attending to others' learning processes, as facilitator and without any stated goal for my own learning, such attention appears to stimulate and facilitate learning that matters greatly to me. I do not know how this happens.

Moreover, it appears that if I abandon my own learning pursuits and my own deep, if ambivalent, investment in them, I function less well as a facilitator.

A parting note on leadership

The trainer—as designated group leader or process facilitator—has a unique role in terms of managing her or his personal problems with group membership, with added responsibility for attending to the troubles of individuals and of the group as a whole. As van Beekum (2012) wryly and wisely noted, "The basic rule of scuba diving applies: Do not take your client[s] to deeper levels than those to which you have had the courage to go yourself" (p. 132).

In fact, another word for professional capacity—whether for deep diving or other forms of work—might be leadership. The group facilitator's capacity for leadership is being developed even as she or he seeks to facilitate that development in the group participants—trainees, supervisees, managers, and community workers, to name some.

In that sense, my leadership emerges from, or functions in synergy with, my willingness also to be a member who is learning with the rest, in company, which I never take as permission to abandon my responsibility to the group. Perhaps this models the tension we all face in terms of what we owe our groups and what we owe ourselves individually.

So when the training process is working, the trainer is also matured. Berne (1966) wrote beautifully of this:

> As the group therapist takes his seat before the assembled patients, his first concern should be to compose his mind for the task which lies before him. He should make a point of starting each new group, and ideally each new meeting, in a fresh frame of mind. It is evident that if he conducts his groups this year just as he did last year he has learned nothing in the meantime and is a mere technician. For the sake of his own development and self-esteem he should not allow such a thing to happen. He may set as a goal (which he may not always be able to attain) to learn something new every week—not something new out of books, nor some interpretive sidelight, but some more general truth which will increase his perceptiveness.
>
> (p. 61)

In that spirit, I too appear gradually to gain in perceptiveness, not always conscious, even as I gain in capacity for deeper dives. The urgency to learn seems insatiably lured into increasingly troubled realms, out of which can emerge, by grace, the skills to satisfy our raw desire for living, in what life remains, whether our own or that yet to come.

6

MATURING AS A COMMUNITY EFFORT

The bind

When I first read the two group conference papers, one by Farhad Dalal (2016) and the other by Andrew Samuels (2016), I thought they made a useful and important pairing. Each paper had originally been delivered as a keynote speech at the 2015 International Transactional Analysis Association conference hosted in Sydney, Australia, which focused on the power of group dynamics. Dalal spoke about what he saw as the twin tyrannies of internalism and individualism, that is, the problems that occur when we focus primarily on internal psychological dynamics with an emphasis on the individual at the expense of a broader understanding of social dynamics. Samuels, in turn, spoke about what he saw as being the importance of the individual's role in the group, particularly in the political sphere, as distinct from collective action, in effect offering a more complex view of individuality.

Although their positions initially appeared diametrically opposed, I saw them as describing different sides of a single phenomenon, one that I would consider central to the human condition: me–them, us–you, or any other combination of pronouns-in-tension. That is, as human beings, we are fundamentally embedded in groups (families, communities) even as we are each an individual, unique instance. The tensions between these two aspects can, at times, be so great as to collapse these otherwise complementary positions into just one or the other—for example, the individual's incorporation into and erasure by the rabid, devouring group or the group's elimination, even murder, by the individual's flight into psychotic omnipotence.

I see violence in the breakdown of the held tension and complementarity.

In my view, Dalal and Samuels articulated these different dimensions with a fullness that does not eradicate the other side. And I believe each dimension deserves its fullest articulation for us to know it most clearly. We need passionate description of the kind these two authors have provided in order to understand each facet in itself even as we recognize the dependency of one on the other. For along with the tensions between group and individual, there also exist possibilities for rich cross-fertilization. Yet whether for good or bad, handled gracefully or not, it appears we have no choice but to live this interdependence, even if we never quite learn how.

For most of my life, as I described in earlier chapters, I have inhabited this me–them/us–you tension primarily as a function of my immaturity. I have been intensely sensitized to the group and to family, watching over others as a means of watching out for signs of rejection and other dangers, monitoring so I can herd others back into familiar and safe-seeming patterns, overfunctioning under the guise of taking up the others' slack while actually trying to run things my way. This immature version of individuality has masqueraded as teamwork at the expense of actually defining any individuality, of sticking out my neck, of growing up, of really belonging. Accompanying this lack of personal development has also been a latent hostility toward my groups, which I explored in earlier chapters: relating to others with paranoid mistrust, cutting off when it has suited my intolerance for anxiety.

No one has benefited from these maneuvers.

Let me hasten to add that mine has not been the only immaturity involved. I and my others have anxiously colluded in not growing up, probably over generations. We have lived the problems that Samuels and Dalal describe. Yet reading their two papers also offered me an opportunity to see what I and others around me have been able to grow toward—the activation of our selves as a function of engagement and actual care of the group, care of one another—with a corresponding capacity both to contribute to family and community and to receive in kind. I consider this a version of the gift economy that Newton (2011b) described in relation to the international transactional analysis community.

Power relations and accountability

One significant impediment to my maturing in relation to my groups—and presumably also an impediment to my group's maturing—has been the great difficulty of facing the mounting evidence of the extremes of human cruelty and destructiveness toward one another, which Dalal (2016, pp. 95–97) discusses in terms of power relations. I would see one aspect of this power as the compulsive force to destroy those in the out-group that courses through the family, organization, and community with an activated ferocity that is sometimes impossible to contain. This would be an example of the violence that erupts when the previously mentioned "held tensions and complementarity" collapse. And I would argue that we know of this kind of violence and terror, if only unconsciously, even when we are not its immediate aim. I believe many of us live this fearsome knowledge day-to-day, sometimes in nightmares. Yet this evidence-based fear is not a delusion, although we may attempt to allay it with delusion: fantasies of warm, well-meaning gatherings; fictions of secured positions within the inner circle. In fact, any securities I do enjoy—of which I enjoy many as an educated, white male in a wealthy, Western economy—are secured only by the probable certainty that the torrent of horrid ugliness will be directed first toward those who have been marginalized, pushed to and over the edges. The target is over there, I can point and say, not here with me or mine. And thus we, the privileged, can sleep serenely in our burrows tonight.

The problem of individual accountability and corrective political action is complicated, as Samuels (2016) discusses in terms of "the limits of individual responsibility" (p. 105). I see his position as endorsing humility in the face of great collective troubles that will not be resolved by any individual's (or group's) fantasy of an idealized rescue or resolution. I think he is also pointing to the need to restrain individual overfunctioning, which can create the illusion that something important is being done when, more likely, we face difficulties that can only be addressed if we each contribute and commit to their resolution over a long, hard passage. We cannot force compliance with any restitutive path, as I attempt when nicely pushing my more passive companions in the direction I think is best. Nor can the resolution spring from the mind of any one genius-savior. We each see and have our part. Yet I can find this a maddening state of affairs: "What do you mean I cannot correct this grievous situation on my solitary own? What do you mean I can only do my part in interdependent interaction with the rest of you?" My frantic mindset reflects something of the strain that accompanies our grouped condition. The question of how to avoid living out the group's tyrannical potential and its embodiment in individual tyrants seems to be at the heart of what both Samuels (2016) and Dalal (2016) have articulated from their different viewpoints.

What does our professional community live?

Dalal (2016) characterizes transactional analysis theory (and, by implication, practice) as embodying both the developmentalist and rationalist (or cognitivist) conceptualizations of the human condition. From the first of these positions, he believes transactional analysts would see "the inherent goodness of the [individual, newborn] seed [as being] contaminated by toxic processes intruding from the environment [over the course of its life]" (p. 90). At the same time, he adds, professionals adopting a Bernean perspective would see life's difficulties as being "due to habituated errors of thinking that, when understood, free the person to think and feel differently" (p. 90). By way of example, Dalal discusses the explanation for racism as a contamination from the Parent ego state, an explanation he criticizes as inadequate in terms of understanding how such destructive power dynamics come "to reside in society at large" (p. 93). That is, although such an interpretation might help me, as an individual, to understand something of my internal, racist psychology and its immediate familial history, it does not give me a clear enough picture of the greater collective force and its legacy, within which I am both implicated as a group member and in relation to which I must define a realistic personal position.

It may be that, in theory and in discourse, transactional analysis does privilege the individualistic at the expense of the collective. However, I would also like to consider my experience of the transactional analysis community and its culture as I have known these for over 15 years, as I have seen and been among transactional analysts living and working.

Naturally, I have witnessed many examples of narrow individualism and bullying politics. We are human, after all. At the same time, I have been the direct recipient of respectful confrontations as well as warm encouragements that have helped me to grow personally and professionally. I have been part of gatherings that have worked hard to include difficult, struggling individuals. I have observed many examples of generosity and prosocial action that run counter to Berne's (1972) more self-serving "existential motto" (p. 295) or "dynamic slogan" (p. 296) about transactional analysts being "'healthy, happy, rich, and brave, and get[ting] to travel all over'" (p. 295). For although I have no training that would certify me as a similarly "rich and brave" insider, I have been welcomed into this community. My aberrance has been appreciated, perhaps in direct proportion to my own appreciation of what this community has taught me. I have also witnessed the development of transactional analysis communities in different parts of the world that have drawn productively on this shared tradition of theory and practice, as if that tradition carried within it the potential to balance the more troubling extremes of individualism and mob action.

All told, I have observed, in this professional community, opportunities for diversity and equality coexisting in sometimes uneasy tension with anxious demands for conformity, control, and hierarchy. So, to see how we actually live this group–individual tension, I believe we need to look not just to our literature but also to what we have actualized in practice and in life.

Maturational potential

In referencing the various metamodels of human psychology and psychotherapeutic practice, Dalal (2016) acknowledges having benefited from what he calls the "growth model" (p. 88), even as he critiques those aspects of the model that overprivilege the sanctity of the one; that posit group life as an impediment to becoming one's truest, purest self; that even frame individual maturing as the individual's innately self-directed capacity or—failing that achievement—her or his fault, which I would summarize playfully as: "I am to blame if I cannot harness my inner, innate superpowers in order to blaze skyward into a kind of demidivinity!" More seriously, though, I think this is also what Samuels (2016) is cautioning against in terms of promising too much:

> If one tries to do *tikkun* [repair and restoration of the world] from too perfect and pristine a self-state, it will not work because the only possible way to approach and engage with a broken and fractured world of which one is a part is, surely, as a broken and fractured, stunted individual, an individual imagined with death in mind.
>
> (p. 105)

We might speak as well of approaching and engaging with the broken and fractured groups of which we are each an inextricable part.

Even bearing in mind this preponderance of imperfection and mortality, still there can be something precious and wondrous about each individual instance of flawed life. At the same time, it would seem that life is more than any single instance. Rather, it is an ongoing process, one requiring metabolization and reproduction, that is, taking in from the outer world in order to survive and then transferring those same life-sustaining biomechanisms into the next generation so that life may continue, even if not for eternity. So, although life must continue past the many beings who have come into and gone out of existence, still those distinct, fleeting lives have also formed the very ground for the life that continues on.

Within the life spans of the many who have come and gone, maturing can certainly be seen as a process for individuals, which reflects the growth model that Dalal critiques. Yet I would claim that such individual cases of maturation are of vital interest not just to the individual but also to the group—from the birth of an infant learning to live with greater autonomy all the way to the growth into wisdom of the adults who may next lead the group toward sustainability. Each individual's maturing happens, in part, as a function of involvement with the group, just as that maturing is also what allows the group to survive over time. I see maturing as a transgenerational endeavor and phenomenon, even given the degree of nonmaturing and nonlearning that we may each witness over a lifetime, sometimes occurring over generations of a family's or community's history.

Although I speak of maturing as a process for the individual and the group, I do not see these as equivalent processes. After all, the biomechanisms that constitute an individual human being—the internal organs, the biochemistry, the neurophysiology—are not what constitute the biological organism or system we call the group or family. By analogy, I would distinguish between the processes that constitute the growth of an individual tree and those that constitute the development of the ecosystem in which that tree finds its life. Yet even acknowledging the differences in actual processes, I have been part of groups and communities, in the human realm, that have clearly grown in shared knowledge over time, knowledge that I have been fortunate and grateful to receive, wisdom that has fostered my personal maturing and that, in turn, has allowed me to foster such growth in those coming after me.

I have lived the sense of something precious having been attained over time, with collective effort, which can sometimes feel as precious and delicate as the next new instance of life.

The fragile arc of aspiration

One concept from transactional analysis that has been of particular importance to my personal and professional growth has been *aspiration*, first introduced as a component of the theory by Berne (1972), later highlighted and elaborated by Clarkson (1992), and then extended further by others in the larger transactional analysis community (e.g., Barrow, 2011; Chandran, 2007; Cornell, 2010; Milnes, 2019; Piccinino, 2018). And for many years, my conceptualization of aspiration has, indeed, privileged the individualistic. Yet over time, in the course of maturing in my likewise maturing communities, I have come to see aspiration as a force that serves both the good of the individual and the good of the group, again as these are held in tension, yet without which tension and reciprocity no degree of aspiration can truly be achieved.

Even this chapter you now read, which exists as the signed work of one individual, was given birth as a function of a vibrant confrontation with a trusted colleague and an engaged exchange with the words of two authors, both nominally from outside the transactional analysis community yet welcomed in for their generative differences and contributions. Moreover, this whole book, in its final form, was strengthened and polished through vigorous give and take with an editorial team. And with fortune, perhaps the words here—owing so much to the group—will, in turn, be plowed back into the group's ongoing life and struggle to live the human condition, especially its pronouns-in-tension.

Such moments of generative potential are often attended by that passion and small miracle we call gratitude.

7

GROUPS THAT LEARN AND GROUPS THAT DON'T

As a child watching groups of people—my family, my classmates, crowds at Mardi Gras, passengers waiting to board an airplane—I had some tacit sense of those clusters as creatures that were somehow greater than the simple sum of their parts, with something more vast in terms of a life force, almost the way a herd of bison, swarming en masse across a dusty plain, can seem like a large, thundering organism. Of course, the people I saw typically offered nothing as dramatic, with the exception of the time in 1978 when I was in Paris watching a march commemorating the ten-year anniversary of the 1968 demonstrations. Quite suddenly, a riot—of violence, screaming, thrown objects, and God knows what else—broke out on the street where I had been standing. The crowd erupted with an almost volcanic force, and I only remember that something instantly moved my body with considerable urgency to flee and find a place of safety.

Even when I later began to study groups more formally, it became increasingly normal for me to relate to them as living entities, as beings in their own right. And not just conceptually. Today I also relate to these collectives emotionally, as I might toward my garden or a beloved family pet; or toward the general welfare of my coworkers, day-to-day, as distinct from my concern for each as individuals; or toward the ecology of a forest system in one of our nearby parks, as the trees, birds, insects, rodents, fungi, and smaller plants interact and adapt over the seasons; or as I even once felt toward the sourdough starter I nurtured for years as essential to my weekly bread baking, as something sustaining my life. These have been entities and relations that have mattered to me. I have felt their meanings and importance.

As the process of learning has become so central to my professional work and to the ways I interact with the people in my life—my family, my community, my workplace, my meditation group, to name a few—I also began to think about groups in terms of their own learning processes, perhaps something that could be understood and cultivated. For although human learning is often framed in terms of the individual's progress or lack thereof—even when conceptualized as a relational process—I wondered whether human groups, as living systems, could be capable of learning in their own way. And if so, could such learning be facilitated? Would it be possible to redirect a group's nonlearning impasses?

A key factor in this quest involved a change from thinking about group interventions in logistical, almost mechanistic terms to appreciating and developing a feel for groups as biological, sensing systems that have to be worked with more organically, as many of us quite naturally do with individual people. Moreover, in writing about groups that learn (and don't), I want to underscore what I see as the importance of learning for all living organisms, among which I would include groups of human beings.

Groups as living, essential systems

In his efforts to expand the conceptions of human psychology then prevalent in his time, Berne (1947) borrowed the ancient Greek term *physis*, which he defined as: "The growth force of nature, which makes organisms evolve into higher forms, embryos develop into adults, sick people get better, and healthy people strive to attain their ideals and grow more mature" (p. 306).

In using this term *physis*, Berne conveyed his sense of a life force that not only animates humans but that represents the growth force of nature for all organisms. He was describing, in my view, his sense of a common attribute among all living things.

Similarly, Bowen (1978/1994)—in conceptualizing the human species—was interested in the

> automatic forces that govern protoplasmic life ... [including] ... the force that biology defines as instinct, reproduction, the automatic activity controlled by the automatic nervous system, subjective emotional and feeling states, and the forces that govern relationship systems ... In broad terms, the emotional system governs the "dance of life" in all living things. It is deep in the phylogenetic past ...
>
> (p. 305)

In that spirit, human groups—in the form of families, organizations, communities, nations, and other human collectives—can be said to function as living, dynamic systems animated by that force within us, that *physis*. And as biological organisms, groups are made of interacting, cross-regulating human bodies that analogically comprise the "cells" and "organs" of each group's body. And those bodily members—in continual, interdependent interaction with one another—create what I referred to earlier in this book as "the shared bodymind," an entity that is simultaneously greater than any single component and yet is constituted precisely from the dynamism of each component, each individual in interplay.

Following Bowen's concept of humans as natural systems, I contend that humans band together and form groups and collectives for many purposes—warmth, safety (in numbers), to fight a common enemy, to share resources and knowledge, to procreate and prepare the next generation.

Yet in conceptualizing the human as part of nature, I do not want to deny the distinguishing characteristics of our species as compared to, say, the distinguishing characteristics of the bacterium or sea cucumber or sparrow. Rather, I want to ask: What traits are common among the different forms of life? What do we share that might help our self-understanding? Mark Johnson and Tim Rohrer (2017), both philosophers of cognitive science, articulated a similar quest for such commonalty among forms of life by drawing on the comparable efforts and thinking of John Dewey and William James to explain "how we move from single-celled animals all the way up to the highest cognitive achievements of humans" (p. 67).

Bowen, in some of his writings from the late 1950s, was likewise trying to develop a psychological formulation of schizophrenia, one that would be grounded in understanding the human as a part of nature. He wrote of wanting to understand "the lawful order between the cell and the psyche" (Bowen, 1995, p. 20)—in other words, to understand how life went from being bacterial all the way to theory making, to see the evolutionary commonalty among forms of life that on the surface appear so different.

His perspective profoundly affected my thinking about human life as being fundamentally shaped by the requirements of metabolism and reproduction that apply to all forms of life. To that I eventually added the essential characteristic of responsiveness to internal and external environments, which became the foundation for my conception of learning.

Seeing this connection among all forms of life has, in turn, helped me to see the concepts of Bowen's theory more clearly—as when I observed the fishes in the Pittsburgh Zoo aquarium and then could see the process that Bowen (1978/1994) called "anxiety" operating in the shoaling and schooling of those fishes. This then allowed me to see that very same system-level movement and responsiveness in my family and in other groups of which I have been a part. Perhaps more so, looking at my human life as just another aspect of life broadly has helped me to be more neutral and compassionate with regard to human limitations that do not live up to the more idealized notion of the human as some brilliant, surefire prodigy, the most dazzling thing to come down the cosmic block, unlike anything that came before or will come after.

Instead, the implication is that we find ourselves, as individuals, inseparable not only from the larger natural environment but also from our human systems, certainly embedded in families but also in groups that allow us to make our lives—to grow food, raise children, help one another, learn the patterns of our world, find meaning, practice spiritually. It is in line with that contention, I believe, that Bion (1959/1969) also spoke of the group as being "essential to the fulfillment of a [person's] mental life" (p, 53), that is, fundamental for our learning and meaning making. As quoted in an earlier chapter, Berne (1963, p. 159) likewise noted the importance of groups for preventing "biologic, psychological and also moral deterioration" with few us of able to "recharge [our] own batteries," lift ourselves by our own "psychological bootstraps," and keep our "morals trimmed without outside assistance."

Bowen (1991), nearing the end of his life, wrote about this quite beautifully:

> Every human infant starts life fully dependent on others, specifically on the family of origin. Growing up involves progressive development of individual characteristics, and aspects of increasing independence. The development of self occurs, in the case of each person, in and through networks of relationships with other members of the family system ... When each member of the family can act as a self, from self, the functioning of the family, as a unitary whole, is improved. But the development, as selves, of the children born into that family depends on the emotional system of that family. Only if the family functions well can the children develop as selves. *Out of unity* (of the family) *comes diversity* (individuality).
>
> (pp. 89–90)

I believe that when we ignore this essential grouped aspect of our existence, we impair our abilities to face and address some of our greatest challenges in family, professional, and community life. We then also fall short of achieving our dearest aspirations.

My conception of learning

For many of us, "learning" can be a pretty worn-out word, one with too many definitions or shades of meaning, possibly a cliché that ends up meaning little. I was interested to discover that the Latin root of "learning" is *lira*—which indicates a furrow or track—learning along a particular groove, learning perhaps a particular straight and narrow. But as I know from my childhood exposure to Spanish, the word for learning in that language is *aprender*, which relates to the English for "to apprehend, to grasp." This implies understanding more than just conforming to a narrow track.

As noted earlier, such apprehending and furrowing is commonly framed as a process within individual minds and bodies. Yet even if we can think of learning as a group process, I believe there is still some usefulness to the emphasis, in standard learning theory, on individual neurophysiology. This framework and field of research can be expanded both to situate the individual's process in the group as well as to distinguish the neurophysiology of the individuals in the group from the neurophysiology of the group as an organism in its own right.

Speaking first at the individual level—and borrowing from Kolb's (1984) notion of "integrative development" (p. 209)—I will define learning as *the neurophysiological integration of the data of experience.* By this I mean the way our bodies—from birth until death—receive, process, and attempt to make sense of signals from the internal and external environments along multiple sensory channels.

We can see this kind of learning from the toddler who figures out how feet can work, and then perhaps how to walk in relation to a world and to other objects, all the way to the university-trained geologist who clocks hundreds of hours observing geographic strata while studying the writings of others who have previously researched and thought about the Earth's history and, in so doing, perhaps coming to some previously unseen conception.

Given the infinite nature of experiential data and the structure of human neurophysiology, this process of reception and integration appears to be a fundamental and lifelong human task.

In using this particular definition, I am not talking about learning as just acquiring new information, although new information is often a result of data gathering. Rather, I am referring to the added step of integrating that new data or information, of making some sense of it, a crucial step that permits further contact with life and further development in the capacity to pursue goals for one's self and one's group. We can think of this also in terms of making meaning of what we encounter.

Although quoted in an earlier chapter, I believe Berne's (1968a) assertion bears repeating here:

> One of the most important things in life is to understand reality and to keep changing our images to correspond to it, for it is our images which determine our actions and feelings, and the more accurate they are the easier it will be for us to attain happiness and stay happy in an ever-changing world where happiness depends in large part on other people.
>
> (p. 46)

This urgency to understand reality was also described by Bion (1962/1967c) in terms of "thinking," which he saw as a mental and bodily drive that is "called into existence to cope with thoughts ... forced on the psyche by the pressure of thoughts and not the other way around" (p. 111). I see these "thoughts" as the not-yet-integrated data or impressions of life experiences. For Bion (1962/1977c), the bodily drive to think in order to know is as powerful as the drives to love and to hate.

So, to borrow Berne's term, my "image" of learning is not primarily about basic stimulus–response conditioning. In Chapter 4, for example, I wrote about social pain and the limitations of the pain-cued learning that can occur in such situations, even though there are occasions when such stimulus–response training can be a useful part of a larger learning effort. Rather, I usually think of conditioning as giving the appearance of learning when, in fact, it more often masks a chronically anxious process: "This is how we always do things (even if it has nothing to do with the actual world)." Thomas Kuhn (1962/2012b) referred to this as "normal science" in his book on scientific revolutions, that is, business as usual.

Finally, I am also not talking about trauma as learning, which by definition indicates the body's incapacity to process the data of certain experiences. In fact, trauma involves an overwhelm of experiential data that interferes with the inborn capacity for learning and integrating. In traumatic conditions, the system is overloaded, unable to move forward, at least momentarily. We see this not just in individuals but in families and communities that have been unable to find ways of coming to terms with certain horrific and painful experiences and histories, sometimes for many generations.

As I hope to show, my proposed definition of learning applies not only to individual human beings but to other living organisms as well, even those with less complex cognitive capacities, such as human groups. Indeed, thinking of groups as living systems—capable of the learning that must occur for all forms of life—is a perspective also shared by organizational consultant Margaret Wheatley (2017). Most importantly, I believe this conceptualization of learning at the group level can be useful whether we are the group's designated leader, an engaged group member, or a consultant to a group's leadership and followership. The question of how that learning can be facilitated—framed in terms of principles and practices for group work—will be covered in a later chapter, especially when attempting to redirect a group's nonlearning impasses.

Groups as living, learning systems

As I wrote earlier, I used to caricature groups as being brainless, which in a sense is true: Groups do not have a brain of the kind we individuals do. But neither are they stupid. Still, I kept asking, if groups do not have brains, per se, how can they be said to learn?

Over time, I realized the biases operating in that statement, one being that only humans are capable of our kind of learning. Another bias is that the best model for cognitive processes is the individual human, which I believe makes it hard to perceive the intelligence and learning potential that groups and, indeed, other living organisms can demonstrate.

In Chapter 3, I wrote about groups responding to threatening environmental cues with a few preprogrammed maneuvers, which Bion (1959/1969) called "the basic assumptions"—to fight, run away, secure the borders, elect an idealized leader for the rest of the passive flock to follow, and so on. I will discuss these dynamics more later. Indeed, there are times when acute anxiety is high in relation to a real and immediate threat and such automatic, emotional maneuvers can be quite adaptive, just as there are times when chronic, delusional anxiety is high and those same routines can be quite maladaptive.

But along with those automatic responses, I also came to see that, when conditions are right, the intelligence of the group can function like a distributed network or system of processors, with problems broken down into components to be solved by the different processors—by the individuals with more complex brains—and then reintegrated at the system level. I have come to believe that something happens to our individual intellectual capacities when we bring our minds together in collaborative ways, something that a single human brain cannot achieve alone.

How might such an understanding matter to me as an individual? Although I see my life as inextricably entangled with the lives of others of my kind—an acknowledgment of my interdependence on the larger ecosystem around me—I do not conclude that the notion of individuality is a useless myth. Rather, I see my individuality as understandable only in its encompassing context, which above all is a transgenerational process in which I have inherited much of my body and mind from those who came before me. And once my limited lifespan has ended, others will likely live to inherit what I have passed along—gifts and liabilities alike. Indeed, I already live this transgenerational concern in terms of my offspring; in terms of the college-aged adults I see for psychotherapy who are young enough to be my grandchildren; in terms of the younger professionals I supervise, teach, and mentor; and in terms of the practices of community care that I seek to foster so that those practices may offer benefit even after I and my age-peers are gone.

So, although groups appear incapable of the kind of cognitive processing seen in individual humans, they do evidence a capacity for processing stimuli from the environment and from within the group itself. They are then capable of using that processing to adjust to changing conditions. As noted, human groups can coordinate activities in the face of external or internal threats, but we also see them achieving more complex tasks, such as producing food and shelter, developing integrated circuits, establishing religious traditions, and cultivating scientific communities, to name a few. This suggests a cognitive-like process at the group level.

To better understand this process and the learning that occurs at the group level, I decided to follow Bowen's interest in understanding "the lawful order between the cell and the psyche." At first, I thought I would learn more about amoebas. But then I found out that the amoeba is actually a rather complex single-celled creature. So I turned, instead, to the simpler and evolutionarily earlier single-celled bacteria.

I also liked the challenge of identifying what I once would have considered to be the height of absurdity: the notion that learning occurs even in bacteria, that there could be a clear, useful link to be found between cell and psyche.

I did not have to look far. In the fifth edition of their classic textbook *Biochemistry*, Berg, Tymoczko, and Stryer (2002) wrote:

> To survive in a changing world, cells evolved mechanisms for adjusting their biochemistry in response to signals indicating environmental change.
>
> (p. 33)

What seems key is the notion of sensing and then responding to environmental change, which can occur in the larger surrounding world as well as within a bacterium's internal biochemical milieu. In an article titled "How Does the Lowly Bacterium Sense Its Environment?," Ramanujan (2006) wrote, "receptors assemble into a kind of cooperative lattice on a bacterium's surface to amplify infinitesimal changes in the environment and kick off processes that lead to specific responses within the cell" (para. 3).

That notion of a "cooperative lattice" fits what Gray (2015) described in her report on the work of researchers at the University of Washington, who found:

> Bacteria can pick up external signals, which then relay to internal signaling pathways that direct their behavior. This surveillance also can trigger survival tactics for a variety of harsh situations, such as lack of nutrients or the presence of antibiotics.
>
> [For instance,] to adapt so readily ... Salmonella typhimurium bacteria need to figure out if they are in the stomach, within cells, or on a plant or other surface.
>
> (Para. 1, 4)

Hopkin (2008) took these varied observations to the next conceptual level in his article "Bacteria 'Can Learn'," drawing primarily on the research of Saeed Tavazoie and colleagues at Princeton University:

> *E. coli* colonies can ... associate higher temperatures (as found in a human mouth, for example) with a lack of oxygen (as found

inside the human gut). When exposed to higher temperatures, they alter their metabolism in anticipation of lowering oxygen levels ...

This result is unexpected, Tavazoie explains. "For as long as people have been studying the behaviour of bacteria, they have assumed that responses to environmental stimuli occur in an action–reaction fashion," he says. This concept, often known as homeostasis, has dominated the field for a century ...

"What we have found is that homeostasis is not the whole story," Tavazoie says. Bacteria can learn to respond in a way that is more complex than a simple reaction to current conditions; it anticipates future conditions ...

It's the first evidence that bacteria have an ability for "associative learning', Tavazoie adds. That's not to say, of course, that single-celled organisms learn in the same way as dogs, or people ...

"Associative learning in dogs and humans happens over the course of the organism's lifetime, and involves modifications to the strength of connections between neurons in the brain," Tavazoie says. "The learning that we have discovered in bacteria occurs over a long evolutionary time-scale and involves changes in the connections between networks of genes."

(para. 3, 5–8)

As I read these descriptions and grappled with unfamiliar biological concepts, the notion of learning began to seem like a process common to all forms of life, one that does not require neurons, for example, or more complex cognitive processes. I also began to think that for humans, learning might also be both a cognitive, neurological process and a more embodied, noncognitive process. These might be working concurrently, maybe even cooperatively, as Bowen conceived of the human bodily emotional system and the human intellectual system sometimes working in tandem and sometimes not. Johnson and Rohrer (2017) proposed a similar interplay that they call an "embodied realism [whereby] thinking is a form of bodily action in the world with which we are in touch through our bodies" (p. 67) and in which "cognition does not take place only within the brain and body of a single individual, but instead is partly constituted by social interactions and relations" (p. 91).

For the moment, consider that there are three requirements for life, at whatever level of complexity: *metabolization* (taking in and making use of materials from the environment); *responsiveness* (adjusting to environmental changes, within realistic limits); and *reproduction* (passing along genetic material to the next generation).

The human organism, as a social species (as bacteria are also now understood to be), certainly meets these three basic requirements, especially when viewed as a somewhat coordinated grouping of individual humans mutually regulating and interdependent, with a capacity for some self-definition and thus more sophisticated, group-level coordination. This human organism, in its grouped form, operates with an emotional system, one that may be as capable of learning as any enduring bacterial colony. I am speaking here of the kind of learning that allows for evolutionary change and survival across generations.

In short, learning is necessary for ongoing life in the changing environment that is our world. To that end, we can say that a bacterial colony is a group that learns, just as the human scientific community is a group that learns, as I will discuss later in this chapter.

Groups that don't learn

Any organism that fails to learn will, in time, fail to survive. And it remains to be seen how our human species will learn, adapt, and survive over the longer span of evolutionary time. For although I believe it is important to identify and attend to the qualities of human groups that can and do learn, I think it equally necessary to bear in mind and body the qualities of our collective nonlearning. After all, humans are also fundamentally inclined to engage in behaviors that impede learning and development. It is, therefore, useful to be aware of those emergences, which we may sometimes sense, in our "guts," before something clear has dawned in consciousness.

In fact, what often prompts us to attention and action is our awareness—or perhaps just our feeling—that a group we are part of is not functioning well. And in our professional roles, we are most likely to be called in to repair those groups, teams, or organizations that are performing poorly.

My three intellectual predecessors and influences—Berne, Bion, and Bowen—have had much to say about these problematic dimensions of group life, conceived respectively in terms of lethal games, destructive basic assumptions, and unregulated emotional processes. I find it interesting how their concepts not only differ but in some cases overlap and correlate with one another, which we might expect given different observational perspectives on the same human phenomenon. What is also interesting is how their views of nonlearning are so intertwined with the potential for learning.

To start with Bowen's (1978/1994) perspective, he wanted to articulate a theory of the human as seen in family life. He saw the family—as well as other human groups—as biological systems functioning with two distinctive aspects that work sometimes in collaboration and integration and sometimes in tension and conflict. One aspect he called the *emotional system*, described earlier in terms of automatic processes that are part of our phylogenetic past, shared in common with other forms of life. This emotional system operates nonconsciously, regulated by what he called anxiety—the necessary biological responsiveness to changing environmental conditions. The other aspect he called the *intellectual system*, which is the human capacity to make use of thinking processes, that is, our cerebral cortices and our abilities to reflect and reason. Both aspects are fundamental and necessary for human life. Thus, Bowen saw optimal thriving to occur when human families and groups made balanced, integrated use of both emotional responsiveness and the capacity for thoughtfulness. He called that balanced use *differentiation of self*, by which he meant both the individual's ability to be part of the system as a separate, functioning self and the family's ability to tolerate and nurture such differences within its living system. He borrowed the term *differentiation* from biology, where it is used to describe cells in the human body, for example, that may start with the same DNA but then differentiate into cells forming muscles, blood, neurons, bodily organs, bones, and so forth depending on where they end up in the system. Just as those differentiated cell types are essential to the life of a body, so too is the human family, as an organism, also served by the differentiation of its members. Bowen (1978/1994) wrote:

> [My] theory postulates two opposing basic life forces. One is a built-in life growth force toward individuality and the differentiation of a separate "self," and the other an equally intense emotional closeness [or togetherness].
>
> (p. 424)

We apparently need that collaboration and tension between individuality and togetherness, between our intellectual abilities and our more automatic bodily responsiveness.

Bowen (1978/1994) also thought that when this balance between the emotional and intellectual systems was absent, what predominated was what he called *chronic anxiety*. This reactivity no longer correlated with actual environmental cues and could spread throughout the system like a wildfire; it was a heightened panic that overwhelms the ability for thoughtfulness. In such conditions, differentiation of self is low; individuality is no longer tolerated. Rather, individuals are subsumed into what is commonly called "groupthink" (see Janis, 1971; Whyte, 1952). And for Bowen, when chronic anxiety predominated in a family, severe symptoms also emerged: substance abuse, physical and sexual abuse, schizophrenia, dissociation, scapegoating, and so on.

So, when a family is unable to regulate its reactivity or anxiety, it can no longer make use of the potential of its different members. It can no longer learn or respond adaptively to what life brings. Groupthink predominates over collective wisdom. And rather than benefiting from the pooling of individual cognitions, the family produces, instead, psychotic thinking, sometimes for generations.

Many of these same human group processes were also described by Bion (1959/1969), albeit from an object relations, psychoanalytic perspective. One of his key concepts was the aforementioned *basic assumptions*, which I believe correlates well with the chronically anxious emotional processes to which Bowen was referring. In Bion's words, "Participation in basic-assumption activity requires no training, experience, or mental development. It is instantaneous, inevitable, and instinctive" (p. 153).

Bion also coined the terms *beta elements* and *alpha function*. As discussed in Chapter 1, beta elements refer to the fragments or bits of experience that are inevitable in life and that have not yet been processed or integrated, absent which we see basic-assumption activity. Alpha function, on the other hand, refers to the human ability to formulate rational, integrated expressions in language, which fits with Bowen's idea of the human intellectual system as essential for families and groups fostering the differentiation of self. And that would describe well what Bion (1959/1969) had in mind when he wrote about the productivity and growth fostering of the "*W* group" or work group:

> "Development" [is] an important function of the *W*[ork] group. It is also one of the respects in which the *W* group differs from the basic-assumption group ... Since the *W* group is concerned with reality, its techniques tend ultimately to be scientific.
>
> (p. 127)

Yet Bion's notion of science is not just intellectual. As discussed in Chapter 1, he saw this kind of orientation to reality as an outgrowth of bodily containing the more distressing aspects of human experience rather than discharging or evacuating that distress. Containing, in Bion's conception, is about managing rather than reacting to the kind of intense, chronically anxious emotional process that Bowen named. The concept of containing may also deepen Berne's notion of the Adult ego state, that is, an ability to reason that may necessarily be accompanied by the capacity for suffering the challenges life brings, especially of the kind that can be so powerfully engulfing in our human groups, families, and communities.

As noted earlier, Berne acknowledged his appreciation for Bion's thinking, at times describing basic assumption or emotional process activity in his own words. For example, Berne (1963) wrote:

> Since ... the first task of a group is to ensure its own survival, all other work tends to be suspended in the face of an external threat, and the group mobilizes its energies to engage in the external group process ... A group in this state may be called a combat group. In dire straits, if the will of the group to survive is strong, the whole membership ... devotes its energies directly or indirectly to fighting the threatening external environment.
>
> (p. 26)

Here Berne linked the notion of a combat group—geared up instinctively to fight for survival—with the abandonment of the group's everyday work purpose, its fundamental reason for existing in times of relative peace. In Bion's terms, basic assumptions then predominate over the productive functioning of the *W* group. In such a context, I see Berne also describing unregulated emotional process at the group level, which fits with that same unregulated, nonthoughtful behavior about which he (quoted earlier) wrote of the human's tendency to "take what he wants when he wants it, and to destroy immediately anything that gets in his way, annoys him, or crosses him up" (Berne, 1947, p. 42). Indeed, humans can often act in such automatic, ugly ways.

Yet Berne (1964) was also attentive to what humans could do more productively and creatively with what he called the capacities for "awareness, spontaneity and intimacy" (p. 178). He saw these potentials emerging when the more fundamental, biological needs of the group were addressed:

> The overriding concern of every healthy group is to survive as long as possible or at least until its task is done. The standards of health for a group, as for an organism, are durability, effectiveness and capacity for full growth. And it is clear that the first requirement for all these is survival.
>
> (Berne, 1963, p. 67)

Berne (1963) was also sensitive to the tension between the group as a whole and the individual, as when he wrote:

> First there should be a model of the group as a whole and then one of the individual human personality. This makes it possible to consider the interplay between them: What such an ideal group does to the personality, and what such ideal personalities do in and to the group.
>
> (p. 53)

He was acknowledging, I believe, the back-and-forth between the group and its members, what Bowen was writing about in terms of differentiation of self and its contribution to the welfare of the family or group over time. Indeed, Berne (1963) himself used the term *differentiation*, albeit in a somewhat different manner, to refer to "(1) The distinction of one class from another, or (2) the distinction of one individual from another in a group imago [or mental representation of the group]" (p. 241). That is, the road to healthy involvement in a group requires distinguishing the individuals, as actual people, from the broader, less-differentiated mass of the group's emotional process. And as he wrote in relation to the staff–patient staff conference (Berne, 1968b), the effort to respectfully acknowledge all those participating as individuals, not as roles, would "stimulate thinking and the organization of thoughts" (p. 286), which is what Bion described as being essential to healthy human life.

Berne thus succinctly described the process of group learning even as he also implicitly acknowledged the conditions for nonlearning.

The transmission of culture

If we think of a group's learning as evolving over time, something is necessarily transmitted to the next generation as part of that process. And what is passed along is likely a mix of the previous generation's hits and misses, its best guesses, some of which may support survival over time, some of which may interfere with it. That transmission is a gift we can both appreciate and reevaluate, for it at least gives us a place to start.

Some of what is transmitted is certainly genetic, along with other particular pieces of information, perhaps procedures or facts, previously achieved knowledge or technologies such as tools or written works. But I think it is important also to consider the transmission of the culture of a family, organization, or community, which is a far more complex phenomenon and not always easy to define.

As noted earlier, I find helpful Berne's (1963) description of culture in groups as "the material, intellectual and social influences which regulate the group work, including the technical culture, the group etiquette and the group character" (p. 239). He spoke of these components also as "a regulating force ... [that gives] ... form to the group cohesion" (p. 239), observing that "the culture influences almost everything that happens in a social aggregation" (p. 110). It was the essential function of this cultural aspect that, for Berne, explained why preserving it would be tantamount to fighting for survival, even to the point of setting aside the group's formally designated activity. From this perspective, culture is a prerequisite for aspiration.

In Chapter 3, I offered my own view of group culture as involving sets of tools or schemas; as connoting an environment or atmosphere, with a particular feel and smell, within which life is fostered over generations; and as a verb for the intentional shaping of life, as in husbandry or farming (see Barrow, 2011).

In that sense, I think of any group's culture as emerging and evolving over time, across generations, in a more organic, less-planned manner. Indeed, the concept of a *multigenerational transmission process* is key to Bowen's family systems theory. He described this process as "programmed behavior" (Bowen, 1978/1994, p. 410) that is "transmitted from generation to generation in a genetic-like pattern" (p. 411). Some of the traumatic consequences of such transmissions have been explored in a theme issue of the *Transactional Analysis Journal* (Stuthridge & Rowland, 2019), which featured an interview with psychoanalyst Maurice Apprey, who introduced the idea of *transgenerational haunting* (Apprey, 2014, 2017).

In terms of the less traumatic forms of transmission or haunting, another perspective I find useful is the classic definition of culture proposed by social psychologist and organizational theorist Edgar Schein (1989/2010):

> The culture of a group can now be defined as a pattern of shared basic assumptions learned by a group as it solved its problems of external adaptation and internal integration, which has worked well enough to be considered valid and, therefore, to be taught to new members as the correct way to perceive, think, and feel in relation to those problems ... Culture formation ... is always, by definition, a striving toward patterning and integration.
>
> (p. 18)
>
> Culture is a product of joint learning leading to shared assumptions about how to perform and relate internally.
>
> (p. 21)

As I read Schein's use of the phrase "shared basic assumptions," I appreciated the echo of Bion's concept, which certainly introduces a note of caution for our study of culture in families, groups, organizations, or communities. When are we, in fact, passing along thoughtless, harmful dogmas? I also thought of Kuhn's (1962/2012b) use of the term *paradigm* to talk about the shared assumptions of a scientific community, which, to reiterate Schein's (1989/2010) words, have "worked well enough to be considered valid and, therefore, to be taught to new members as the correct way to perceive, think, and feel" (p. 18). In relation to scientific problems and questions, Kuhn (1962/2012b) wrote that "paradigms guide research by direct modeling as well as through abstracted rules" (p. 48). In that sense, he talked about "shifting emphasis from the cognitive to the normative functions of paradigms" (p. 108), that is, to consider how a paradigm is more than just a set of ideas but also something that shapes and reinforces normative behaviors and perspectives.

I take this also to mean that paradigms, in Kuhn's definition, exist not just for the highly technical realm of science but also for everyday human emotional processes, for ordinary people who operate with compelling, consensual views of reality. Indeed, Kuhn (1962/2012b) wrote about paradigms as being more than just "constitutive of science ... [but] ... constitutive of nature as well" (p. 110), that is, constitutive of human groups as natural systems.

Therefore, I believe we can usefully think of culture as a paradigm, one derived over time through trial and error and that serves to organize human activity on the basis of its presumed efficacy in ensuring the survival of the human family or group. It would also have to be relatively easy to transmit, although that alone would not guarantee its validity or utility. Culture can transfer nonlearning as well as learning.

Learning and the scientific method

When Berne emphasized the importance of understanding reality and changing our images to correspond to it, I believe he was describing, quite succinctly, a key aspect of the scientific method. That is, we gather data; form hypotheses; and then gather further data to revise those hypotheses, ad infinitum, until death or extinction. In other words, what we know with certainty—our latest hypothesis or theory—is always open to new experiential data, to new understandings and potential revision. I characterize this data as "experiential" to encompass what I see as the far greater range of information available to us than simply that which is derived experimentally and deemed, perhaps, more objective. I believe we have to account, above all, for our subjective, experiential data in order to understand reality.

At the same time, Bion (1967b) calls us to consider that, "The psychoanalyst's tool is an attitude of philosophic doubt; [and] to preserve such 'doubt' is of first importance" (p. 157). With this notion of *philosophic doubt*, Bion reminds us that, at the very moment of realization and insight, we can also remain aware that our formulation is almost certainly subject to change. For although what we know now may work well enough for the present moment, our sense of knowing will also need to evolve based on new information and experiences.

The idea of philosophic doubt also suggests a need for an attitude of skepticism, which I wrote about in an earlier paper (Landaiche, 2007) in connection with Berne's recommendations to practitioners:

> Skepticism can be characterized by an emptying of one's mind of certainty and by seeking with curiosity. Berne (1972) suggested that this could be achieved by listening with a mind freed "of outside preoccupations," while putting aside "all Parental prejudices and feelings, including the need to help," and "all preconceptions about [our] patients in general and about the particular patient [we are] listening to" (p. 322). In more colorful terms, he wrote about approaching our work from "the Martian viewpoint," which he defined as "the naivest possible frame of mind for observing earthly phenomena, leaving the intellect free for inquiry without the distraction of preconceptions" (Berne, 1966, p. 366), as if we were looking at this world with the eyes of an inquisitive, possibly more objective outsider.
>
> (p. 27)

This is the stance that Berne (1966) also advocated when, as noted earlier, he wrote about the group therapist composing her or his mind for the work that lies ahead and "starting each new group, and ideally each new meeting, in a fresh frame of mind ... to learn something new every week ... [in order to] increase his [or her] perceptiveness" (p. 61).

I certainly do not come to the scientific method as a scientist or as someone with many quantitative skills. Rather, my approach is from the humanities, in which science, from the Western tradition, emerged as an outgrowth of philosophy—the ethical quest for truth. My conception of science, therefore, covers more than just the technical approaches common to, say, astronomy, physics, chemistry, biology, or engineering. It also includes the more qualitative approaches to researching and data analysis, a perspective of mine largely influenced by those in my professional community who have studied phenomenological psychology, especially here in Pittsburgh where I live (see also Grégoire, 2015). Theirs is the approach to research that I found in Sardello's (1971) notion, previously quoted, of sustaining "an attitude of respectful openness to the whole of our existence" (p. 64).

In this spirit, I think of the scientific method as, quite simply, the gathering of data (through any number of means) and the forming and revising of hypotheses (which eventually become theories)—a kind of "involvement in the world [that allows] reality to reveal itself" (Sardello, 1971, p. 64).

Yet it is not clear which comes first: data gathering or hypothesizing. In fact, they are fundamentally interrelated. We need data with which to form hypotheses, yet data gathering is guided to some extent by already-existing hypotheses or theories. And this iterative, cyclical process appears to result, over time, in the revising of the hypotheses and theories, which in turn modifies the intentional gathering of data. Kuhn (1962/2012b) wrote of these as the paradigms or models of normal science that are in operation until the anomalies that emerge can become retheorized and then researched more closely within a new paradigm or mindset that replaces the old.

The ethos of the scientific method also calls for external or peer review of methods and results, the sharing of data and results, and the replication of results. All of these will bear on my later discussion of the learning community.

For now, I want to focus on research and hypothesis formation. This approach was endorsed by Bowen in regard to observing and continually updating his theory of families. He even cautioned his own followers about reifying what they'd learned from him. In a January 1989 memo to his Georgetown center's faculty and staff, written the year before his death, he expressed it this way:

> A THEORY IS NEVER ABSOLUTE OR COMPLETE. An erosion [of theory] begins when a busy practitioner ... begins to treat theory as a PROVEN FACT, rather than a living thing which is constantly changing. When THE THEORY is considered to be unchangeable, it is automatic for both theory and practice to be distorted.
>
> (Bowen, 1989)

In a presentation I gave in 2015, I described my process of observing, over a period of seven years, the Western Pennsylvania Family Center's Basic Seminar in Bowen Theory. In that presentation, I noted:

> Though we teach Bowen's theory as a particular content, a set of hypotheses fixed at the time of his death, we also facilitate the use of his research methodology, which is intended eventually to revise the content—the theory. As such I believe that what we learn most importantly when we study Bowen's theory is Bowen's *methodology*. We learn the process *behind* the theory's content.
>
> So I will define learning, maturing, and integrating most broadly as the intention to research, to gather data through the lens of the existing theory, with an eye to theorizing further, that is, to revising the initial theory, a learning outcome that correlates with improved functioning in and adaptation to life.
>
> (Landaiche, 2015, n.p.)

Although science can be equated with a kind of absolute knowledge and rigidity, I suggest that such absolutes are not actually the aspirational expression of science. Rather, they are more likely the understandable result of our human tendencies to reify and assuage our anxieties in the face of the vast unknown qualities of our universe. When we are able to manage those anxieties, we can return to learning, to integrating further experiential data about our worlds, and to passing those findings along for future generations. Indeed, I would equate human learning with the scientific method, just as I would equate nonlearning with a more anxious approach to or attitude about science.

So, along with seeing learning as a natural, inbuilt process for all living organisms, I would say the same of the scientific method. As I wrote earlier, the physiological reception and integration of the data of experience appear to be necessary for suitable responsiveness in an organism of any complexity. And in the case of living things, this integration represents a hypothesis or theory about what it takes to remain alive. A bacterium, to survive, must integrate (i.e., process in an organized, practical, reality-oriented manner) the data of experience as an ongoing fact of its life span; so must a bacterial colony. In each case—for the individual and the group—the learned response represents a hypothesis that, if accurate, allows ongoing life.

Moreover, such responsiveness appears to be learned as an outcome of a transgenerational process. For bacteria, the individual bacterium is a product of evolution—representing the selection of the bacterial colony's fittest—and the bacterial colony represents the evolved capacity to cooperate given certain environmental conditions.

Presumably, the more complex the system—the greater the number of moving parts—the more complex the integration process. Certainly, human integration involves such possibly more complex activities as the making of tools and meanings that may facilitate ongoing responsiveness and future efforts at integrating. In the words of James Marcum (2015), a philosopher in the medical humanities program at Baylor University, "The world is ... a niche and science helps a community adapt to it" (p. 142).

Data gathering in science is often equated with quantifiable datum with the use of instruments. Surely these have their important place in different fields of science. But in this book, I have described the simple human act of observing to be a form of data gathering, one that strives also to make use of all our bodily sensations. These sense impressions become the basis for our making some organized sense or meaning from our life experiences. And I believe it is a natural process that we can discipline ourselves to use in a manner that supports the lives of our families, groups, organizations, and communities.

I think we can also see the interconnection among researching, theory making, and differentiation of self—living our individuality as part of a human system or group—in this statement from Bowen (1978/1994):

> The process of differentiating a self ... requires knowledge of the function of emotional systems in all families [that is, knowledge of a theory, of a set of interrelated, generalizable concepts] and the motivation to do a research study on one's own family.
>
> (p. 539)

Thus, studying ourselves in collective life is a living science, one we hope will support our integration more than our anxious, harmful nonlearning.

8

THE LEARNING COMMUNITY

In recent years, I have been using the concept of the *learning community* as a way of thinking about groups that learn. I have been especially interested in what happens when humans think collectively, that is, as differentiated individuals within a cooperative human system (Landaiche, 2016a).

I define a learning community as any human grouping whose primary objective is the support of and engagement with the ongoing process that is human learning. It facilitates learning for its individual members as well as for the group as a whole—the living human system—whatever the focus of that learning may be.

The scientific community would be a prime example of a learning community, one engaged in ongoing research and theorizing, perhaps toward application. The community plays an active role in verifying, validating, perhaps replicating, and surely sharpening the learning of individual scientists while creating a body of knowledge, theory, and methodologies available to all scientists, a foundation on which further discoveries and learning can be built.

Additional examples include sports and other teams oriented toward mastery of certain physical skills, ongoing reading and study groups, learning organizations that rely on continually developing knowledge, gardening and other clubs interested in honing and advancing proficiencies, and food-producing collectives that progress in aptitude and productivity. I also include religious and spiritual communities that seek to develop disciplines and perspectives in order to know or point toward an "ultimate reality," that is, toward an understanding of life as that affects interpersonal behavior and ethics as well as the striving to establish meaning and to reduce harm toward what lives.

The most vivid example of a learning community is one devoted principally to learning about the process of human learning, with the application of that learning subsequently returned to the ongoing effort as a form of more sophisticated researching and theorizing about human learning and human being. In this process, there is a need to learn about all that is human (arts, literature, psychology, history, sociology, human development, medicine, economics, and so on), as I strived to do in my doctoral interdisciplinary studies in the humanities. A collaborative group working toward this kind of human understanding is a living laboratory of and for experiential learning.

Characteristics of the learning community

I see the learning community as engaging in an organic and natural process of ongoing efforts to grow and mature. What can be learned is essentially infinite. At the process level, the learning is both intrapersonal and interpersonal. Moreover, there is a parallel between the development of the individual's neurophysiology (mind–body integration) and the group's development of a conducive environment—stimulating, aspirational, organizing—along with the creation of a knowledge base and culture for facilitating future learning. Cornell (2008) wrote about this in terms of establishing a community that cultivates thinking. Bion (1962/1967c) used the same designation, as did my mentor, Elizabeth Minnich (2003), who also spoke of the intellectual, moral, and political importance of teaching thinking to our young adults.

In these many ways, I think of the learning community as an ecosystem, one that learns over time to make use of human maturing and aspirations. Thus, there is an ecology to a well-functioning learning community, a diverse mix of interdependent, interacting factors that are capable of reproducing themselves and relaying integrative accomplishments across generations. Again, Cornell (2019) wrote about this in terms of fostering freedom for play, imagination, and uncertainty, with his focus being professional learning environments. As noted earlier, Berne spoke of the importance of preserving the group culture. He also believed that "if the group is to survive as an effective force, what must be preserved is … the organizational structure" (Berne, 1963, p. 67). This is, in effect, the structural or institutional means (Douglas, 1986) by which the group, organization, or family has figured out how to coordinate its essential activities.

As mentioned earlier, pursuing authentic, individual, self-actualizing goals appears to enhance neurocognitive development for both self and system. As such, I believe a learning community embodies an ethical stance of concern for individual members, for the group as a whole, and for future generations.

In thinking about these characteristics of learning communities, I was intrigued by what Kuhn (1969/2012a) wrote in regard to the scientific community. He saw it as consisting of

> the practitioners of a scientific specialty … [It involves those who] have undergone similar educations and professional initiations; in the process they have absorbed the same technical literature and drawn

many of the same lessons from it. Usually the boundaries of that standard literature mark the limits of a scientific subject matter, and each community ordinarily has a subject matter of its own.

(p. 176)

A paradigm is what the members of a scientific community share, *and*, conversely, a scientific community consists of men [and women] who share a paradigm.

(p. 175)

We could apply these same descriptors to all types of learning communities, particularly the sharing of a paradigm, that is, an organized view of the world and universe in which we find ourselves. I found particularly interesting Kuhn's (1969/2012a) contention that, if he were rewriting his classic text on the structure of scientific revolutions, "it would ... open with a discussion of the *community structure* of science" (p. 175, emphasis added).

In other words, over and above the content of knowledge and learning, we first want to understand the indispensable structure of the community involved in that content's creation.

Learning in the tango community

To give an example of a learning community, and to take a break from discussing science, I will spend a moment describing tango, a topic that my colleague Laurie Hawkes in the transactional analysis community has taught and written about for over 15 years (Hawkes, 2003, 2019).

Tango is an interesting human invention. The dance requires that one find and remain with one's own center of gravity while dancing with a partner, in fact, while dancing sequentially with multiple partners of varying body types and personalities. The dance also requires the development of a *forward intention* that both signals a lead given and a lead to follow. This involves a leaning onto and into one another while remaining centered, which both supports the dancing pair and gives an indication of each individual's personality and desire.

The leadership and followership roles are defined as reciprocating, with the follower allowing herself or himself to be led (while remaining centered) and the lead recognizing that the follower's center of gravity has to be respected (in effect followed). Not surprisingly, these roles are often gender defined, perhaps reflecting sometimes oppressive gender norms and power imbalances. Yet often dancers learn both roles, and skilled dancing pairs can signal a shift, within the dance, in terms of who leads and who follows, sometimes multiple times within a single dance.

The tango community has passed, and passes, along a shared technology for a dance that is basically improvised within a facilitating structure. The community also establishes a simple "line of dance," that is, a set direction for circling the dance floor, thus facilitating multiple pairs dancing at once. The group also establishes the expectation that one will dance with a given partner for about three or four dances and then switch to a new partner.

For five years in the early 2000s, I was part of the tango community in my hometown of Pittsburgh. Talk about a miserable learning experience! I was clumsy and had trouble getting the basic steps, much less any technically complex maneuvers. The improvisational aspect sounded good in theory but was rather demoralizing. And keeping track of my own center of gravity? Staying aware of it as separate from the person with whom I was dancing and who meanwhile had to mind her own center of gravity? I guess you could call that undifferentiation of self! For without those separated centers of gravity, no couple can dance well. And without multiple couples, centered within themselves, dancing in the same directional circle, aware of themselves and of the others, the whole group cannot dance. As I describe this, it sounds like an impossible mess. But it actually worked. And because it often worked, I kept at it. And every once in a while, I even enjoyed myself.

As we gathered at weekly *practicas* (which were more informal practices) and at occasional *milongas* (which were actual dances, usually dressier), I found the community interaction to be as important as the dance itself. I enjoyed the wide range of people involved. I learned a means of communicating through the dance that did not require much small talk. And I discovered in that embodied communication a sense of depth and satisfaction. Moreover, even beyond that interaction with the community and the dance, I found that the process of developing my forward intention and locating my center of gravity had a powerful effect on my sense of presence when, for example, giving presentations. In leaning ever-so-slightly forward, as in tango, securely anchored to the earth, I reversed a lifelong habit of pulling back, of hesitating, of feeling frightened and voiceless.

I still wonder: What were the characteristics of the tango community, as I knew it, that allowed it to function so effectively, that allowed us to learn not only to dance but also to find our own centers while in contact with multiple bodies? What was the group's modus of transgenerational transmission?

Though these questions require much more intentioned observational research, I think first about the role of culture in facilitating learning for groups. In the tango community, this culture is one of discovering how to use one's own body not just to conform to convention but also to release freer expression through dance, through contact with music that can convey so many passionate rhythms and emotions. Tango offers a form of play that can inform our constructive participation in the broader world.

The learning organization

As I have been discussing the learning community, readers familiar with the organization development literature may also have been thinking of the notion of the *learning organization* popularized by systems scientist Peter Senge (1990). This refers to an organization that facilitates the continuous learning of its members, which, in turn, transforms the organization and its capabilities.

Senge and his colleagues (2000/2012) also wrote about "schools that learn," a phenomenon regrettably not as common as one might hope. These schools, which I think of as learning communities, are described as complex systems that require an "understanding of the relationship among educators, schools, learners, and communities" (p. 5) with the implication that

> institutions of learning can be designed and run as learning organizations ... This means involving everyone in the system in expressing their aspirations, building their awareness, and developing their capabilities together. In a school that learns, people who traditionally may have been suspicious of one another—parents and teachers, educators and local businesspeople, administrators and union members, people inside and outside the school walls, students and adults—recognize their common stake in each other's future and the future of their community.
>
> (p. 5)

Indeed, Senge and his colleagues (2000/2012) offer a broad, systemic perspective:

> Schools don't exist in isolation ... Sustainable communities need viable schools for all their children and learning opportunities for all their adults. In our view, a learning school is not so much a distinct and discrete place (for it may not stay in one building or facility) as a living system for learning—one dedicated to the idea that all those involved with it, individually and together, will be continually enhancing and expanding their awareness and capabilities.
>
> (p. 7)

Toward that goal of organizational and community learning, Senge and his colleagues (2000/2012) proposed a set of disciplines that involve "articulating individual and collective aspirations" (p. 7), engagement in "reflective thinking and generative conversation" (p. 8), and "recognizing and managing complexity ... [that is,] Systems Thinking" (p. 8). I find these disciplines to be quite consistent with the principles and practices I will discuss in the next chapter.

The role of language in the life of our groups

When working with human groups, I have emphasized the importance of attending to many forms of communication, including nonverbal transmissions. This was also something Berne underscored in his instructions to group leaders (e.g., see Berne, 1966, pp. 65–71). At the same time, I believe it is also important to attend to the words and symbols people use, by which I mean the explicit content or meaning of language, the degree to which language feels authentic versus stereotyped, and the extent of freedom allowed in the quest for expression and meaning. Just as Bion (1959/1969) observed: "Verbal exchange is a function of the work group. The more the group corresponds with the basic-assumption group the less it makes any rational use of verbal communication" (p. 185). And to expand on Bion's idea of "verbal exchange," I prefer the broader term, "language," as a way to recognize not just what is spoken or written but what may also be signed with hands, sung with voice, embossed in braille, or otherwise conveyed from one human to another.

As a humanist and writer—as someone who listens to talking all day in my role as psychotherapist and supervisor—language has been an essential part of my life and work. But I did not have a way of thinking about it in terms of natural systems and human groups until I attended a presentation by Yolle-Guida Dervil at the 2015 International Conference on Bowen Theory in Pittsburgh. There she talked about the work of Thomas Schur, a therapist and training supervisor who makes use of Bowen theory and, in particular, emphasizes the function of human language and conversation in the context of natural human systems. In one of his early published articles, Schur (2002) wrote particularly about the supervisor's (or facilitator's) task of managing her or his own reactivity and developing a more solid sense of self, which shows up in the way language is used in supervision or facilitation. Adopting a radically systemic perspective, he contended that, "Since the supervisor is assumed to be part of the system he or she is observing, the conversations between supervisor and supervisee are part of the system of the therapy between supervisee and client" (p. 413). As such, "The supervisor needs to understand more about his or her reactivity and experiment with *changing self in the conversation*" (p. 414, emphasis added). He explained:

> To find the best adaptations to the flow of the system requires experimentation. Each one, supervisor, supervisee, and client, experiments to discover what works, as one is increasingly able to distinguish the old, reactive patterns and able to try new ones in a more thoughtful way. Again, language is the primary vehicle for this experimentation in a recursive flow of thinking and doing that itself is part of the systemic dynamics of the relationship between supervisor and supervisee.
>
> (p. 406)

Perhaps Schur's (2002) most radical position was that he saw "language more as neurological coordination than as delivery of information" (p. 403). This makes sense to me as someone who listens to the way that people use words as much as to their content, the way word usage reflects a stage of neurophysiological development. Indeed, the link that Schur made between language and neurological coordination fits what I have observed in the ways that individuals who research themselves in their systems are gradually able to speak with more authenticity. In that sense, as Schur (2011) put it in a later publication, "one can assess differentiation in a person's use of language, both in reflection and in conversation" (p. 290).

In my own observations of learning (and nonlearning) group situations, I have found it useful to assess not only individuals' uses of language but also the group's overall use of language in both dialogue and debate. As I noted in Chapter 5, I have been particularly interested in moments when conversations among group members appear to be productive, which Shaw (2002) called "conversing as organizing" (p. 11).

Shaw works in the organizational field as a researcher, theorist, professor, and consultant. She takes the radical position that human organizations are the outcome of human conversing. In her view, organizations are not their formal structures, per se, but the way language is used to achieve (or not) their organizational purposes. She is not talking about language as static (or sacred) text, although that would be part of the process. Rather, she is emphasizing the importance of human conversing as continually recreating and reevaluating the organization and its function, which I consider to be a form of hypothesis formation and revision, not to mention essential to creativity.

Shaw's work has been placed under the umbrella category of *dialogic organization development*, which was coined by researchers Gervase Bushe and Robert Marshak (2009) to distinguish an approach to organization development (OD) that they believed differs significantly from what they called classic diagnostic organization development. They suggest that these latter forms of OD are characterized by quantification, objectivity, and the privileging of idealized forms of organizing. While acknowledging a debt to those earlier forms of OD in terms of their contribution to the field, Bushe and Marshak also find those approaches limited. As a corrective, they identified what they saw as the shared characteristics of dialogic OD practices:

- The change process emphasizes changing the conversations that normally take place in the system.
- The purpose of inquiry is to surface, legitimate, and/or learn from the variety of perspectives, cultures, and/or narratives in the system.

- The change process results in new images, narratives, texts, and socially constructed realities that affect how people think and act.
- The change process is consistent with traditional organization development values of collaboration, free and informed choice, and capacity building in the client system.

(Bushe & Marshak, 2009, p. 362)

Of course, there is talk and then there is talk. Some of what comes out of our mouths, especially in groups, is anything but constructive. Ralph Stacey (2001), another organizational theorist considered part of the dialogic OD contingent, observed more neutrally that "communicative interaction simultaneously produces both emergent collaboration and novelty, as well as sterile repetition, disruption and destruction" (p. 148). He then added:

> Any organizational change, any new knowledge creation, is by definition a shift in patterns of communicative interaction, hence a shift in power relations and, therefore, a change in the patterns of inclusion and exclusion. *Anxiety is thus an inevitable companion of change and creativity and so, it follows, are destructive interruptions in communication.*
>
> (p. 156)

I think Stacey was reminding us of the continual tension to be found in group life between creativity and deadness, some of which appears to be reflected in our uses of language that Schur believes can be analyzed to assess levels of differentiation of self.

Bushe and Marshak (2014) sounded a similar note of caution:

> Simply engaging in good dialogues, in creating spaces where people are willing and able to speak their minds, and where people are willing and able to listen carefully to one another, is not sufficient for transformational change to occur.
>
> (p. 77)

They (Bushe & Marshak, 2014) emphasized that the successful use of dialogic OD methods must involve

> A DISRUPTION IN THE ONGOING SOCIAL CONSTRUCTION OF REALITY ... [in which] the previous order or pattern of social relations is pulled apart and there is little chance of going back to the way things were ... A CHANGE TO ONE OR MORE CORE NARRATIVES ... that explain and bring coherence to our organizational lives ... [and] A GENERATIVE IMAGE IS INTRODUCED

OR SURFACES THAT PROVIDES NEW AND COMPELLING ALTERNATIVES FOR THINKING AND ACTING ... [That is, some] combination of words, pictures or other symbolic media that provide[s] new ways of thinking about social and organizational reality.
(pp. 78–79)

To my ear, these sound very much like the features Kuhn described for the evolution of scientific paradigms, which in transactional analysis terms we might also think of as gradually changing an organization's, community's, or family's script.

In my observation of learning groups, I find that learning requires freedom of authentic expression, typically in the form of speech (although not restricted to that, of course). The function of such speech is more on the side of communication than discharge, on the side of clarification more than concealment or subterfuge. Such free, authentic expression represents a state of neurophysiological integration for the individual speaking as well as for the group that provides the systemic opportunity for the dialogue needed for concept development, fact verification, and memory. And although a group cannot, strictly speaking, write a paper, the contributions of individuals in the group can be collated to produce a document with an intelligence or contribution that exceeds that of any one person.

In one professional development group that I observed a few years ago, I had the sense of the group as a body that, in a short time, structured the working space through conversing. The genius of what I can only describe as the group's spoken reverie delivered complexity in a compact form that acted on our bodies, almost as Rolfing does to break through the tightness of old, held patterns so that energy can flow more freely. In that same way, it seemed the group could press onto and remake our individual human bodies in the direction of making greater, collective sense.

I also saw that particular group's body as a mansion with many rooms, enclosures, ateliers, and thematic areas that elaborated and showed a fuller picture of reality, our purpose in gathering, the work that lay before us, and solutions. There was a division of labor and freedom to pursue new work, withdraw, practice and yet still be part of the group—as in those families, organizations, and communities that are committed to learning. The boundaries in that particular group permitted possible growth versus traumatic experiences, such as constriction or chaos. And as a means of dealing with the traumatic material or knowledge that did emerge, there was a metabolization of the languaged experience rather than dissociation or cutoff. There was an ability to use words to link to self-expression and to reach out toward others; there was the production of spoken language that when received by others served to enhance their ability for more authentic expression. The act of speaking did not have to be used to rupture contact.

In groups we may learn to allow this kind of natural process—of soil, fertilization, sunshine, tears. We may find an ancient sense of comfort (of togetherness, safety, peace, and haven) as well as an ancient sense of threat (of being killed). There seems to be little gray zone between these. So how can we trust, for example, that the disturbance in the learning group is safe, is part of the collective good when it triggers that primordial sense of menace?

Harmonious, conflict-free interaction is not necessarily a measure of group effectiveness in regard to the facilitation of learning. Rather, turbulence, at increasingly tolerable levels, seems requisite for the cross-generational process of learning, development, and maturation (Landaiche, 2013). And as Kuhn (1969/2012a) wrote in the final words of his postscript to his classic text, "Scientific knowledge, like language, is intrinsically the common property of a group or else nothing at all. To understand it we shall need to know the special characteristics of the groups that create and use it" (p. 208).

The importance of attending to the human capacity for speech, for using language and knowledge, will always be a key component of the principles and methods when facilitating learning in groups.

9
PRINCIPLES AND PRACTICES OF GROUP WORK

How do we embody our learning and maturing in the work we do within the many groups that form our lives? What are the many subtle ways we show up or not, with or without awareness? Which of those aspects work congruently? Which live a mixed, maybe conflictual message? These are the questions I ask myself often, whether I am a group's designated leader or simply a member. And I think this bears on the interdependence among principles and practices, which I will explain shortly.

As noted in Chapter 5, Berne (1966) wrote about not practicing today as one did yesterday, of starting each new group and meeting "in a fresh frame of mind ... [with the goal] ... to learn something new every week" (p. 61). It is a paradoxical instruction in many ways. After all, we do have to approach our groups with some conceptualization, some fixed ideas. And yet Berne speaks to the continually open process that I think of as being the study of ourselves in collective life. So, toward that goal of openness, I find it helpful to articulate our operating hypotheses, the ideas with which we enter our groups. What are the assumptions we make to start? They certainly guide what we do. And they will almost certainly need to be modified as learning proceeds. So please take what I write here in that spirit of revisability!

In addition to sharing my own guiding hypotheses about learning in groups, I also want to emphasize and elaborate what I, and certainly many others, see as the important, reciprocal group functions of leadership and followership.

I find such reciprocal interaction also occurring between the principles and practices employed in any group that learns. That is, we need practical techniques to do our group work. But those techniques will ideally be derived from principles based on experience and continually evolving as learning proceeds, similar to Lewin's (1948) notion of *action research*. The principles give us a theoretical basis for our actions, even as the outcomes of those actions put those principles to the test and result in their gradual adjustment. I also find it important that the word *principles* connotes values, the things we hold most dear, which I believe keeps our work close to the heart of what matters most to us as individuals and collectives. And yet we have to put our words, our vaunted values, into action; we have to live them to be congruent and honest with ourselves and with one another.

My guiding hypotheses

Given the rather dry feel of the word *hypotheses*, I instead want to begin here with George Eliot's (1860/1961) more rapturous characterization of human understanding:

> All people of broad, strong sense ... early discern that the mysterious complexity of our life is not to be embraced by maxims, and that to lace ourselves up in formulas of that sort is to repress all the divine promptings and inspirations that spring from growing insight and sympathy ... [This is] the insight that comes ... from a life vivid and intense enough to have created a wide fellow-feeling with all that is human.
>
> (pp. 483–484)

Eliot is describing, in my view, the process of studying ourselves in collective life, of researching to make sense of our human condition. For though this "mysterious complexity of our life" often exceeds any easy formulas or maxims, still I believe we each move into our various groups—families, teams, organizations, and communities—with some degree of felt understanding, an embodied experience that probably exceeds our conscious grasp of language. Can we allow ourselves to remain open to that vividness and intensity? Can we learn to trust the natural capacity for integration that is partly achieved in community?

To facilitate group processes, and especially learning in groups, I believe we each draw on the sense we have made and the insights or hypotheses we have constructed to understand human life and human becoming, and these inform our principles and practices. Based on my own experiences in groups, especially these past 30 years of more intensive study and application, I here summarize my guiding hypotheses.

1. **The central human task—from birth to death—is the neurophysiological integration of the data of experience (i.e., learning) as we each move toward our aspirations.**
    - As described in earlier chapters, our bodies are capable of registering, organizing, and conveying multiple modes of information about ourselves and the world of which we are a part. Thus, the

data of experience may be facts, external or internal, and/or information loaded with emotion or emotional implication.
- Nonlearning, for humans, is neurophysiological activity that results in an inability to integrate the data of experience, therefore an inability to develop or mature. At a group or community level, this would be seen as "societal regression," to use Bowen's phrase (1978/1994, pp. 269–282).
- Thus, psychological and emotional symptoms are a function of an impasse in the system's capacity to process or integrate the data of experience, which would characterize Bion's notion of a basic assumption group and Bowen's concept of a family's chronically anxious emotional process.

2. **We achieve integration or maturation through researching, that is, through the scientific method.**
    - As I have written previously in this book, researching and the scientific method involve gathering data, forming hypotheses, testing those hypotheses through further data gathering, then revising the hypotheses, ad infinitum.
    - For humans, neurophysiological integration involves hypothesis formation and theory development, which I observe to be partly an intentional, conscious process and partly an unconscious, more organic process.
    - Researching and the scientific method, although naturally occurring processes, are also amenable to discipline, which Berne, Bion, and Bowen all endorsed in terms of learning to work productively in groups, families, organizations, and communities.

3. **Our neurophysiological integration—learning, maturing—is both an individual and a collective process.**
    - Humans are a social species, with cross-regulating bodies and minds, a condition of our lives that we are wise always to remember and take into consideration when observing our individual behaviors.
    - Maturation is a collective, relational process that can be facilitated (as well as impeded), wherein lies our cautionary hope.

These conceptualizations about human learning guide my efforts at facilitating learning in and for groups. That is, I attend in every moment to evidence of integrating, which shows up as meanings being made, as reductions in chronically anxious behaviors, and as capacities for more effective collaboration toward meaningful goals. Those goals, I believe, are necessarily also consistent with the values of the individuals and of the group. I attend, as well, to evidence of fragmentation, dissociation, overwhelming affects, psychotic notions, and other emotional, psychological, and behavioral indications that integration is not occurring. Such signs of nonintegration are what I find to be most important when I am in the role of facilitating a group process for the purpose of learning toward meaningful work (Landaiche, 2013). I also strive to identify them in myself when I am a group member.

As an example of integrating and learning, I can again describe some of what I observed over seven years in a seminar set up to teach Bowen's family systems theory. In this group, the effort to learn a particular theoretical perspective was based largely on experiential learning, that is, on the study of oneself in one's own family system along with presenting, to the larger seminar group, that research effort.

> As I observed the participants in the Basic Seminar [presenting their family systems], I was struck by the neutrality that developed over time, the way people were gradually able to make contact with family members, as well as family information and dynamics that were previously untouchable. I was struck by the emergence of curiosity, compassion, and humor. Moreover, I noticed that this level of organization appeared to increase in spite of the corresponding increase in the sheer amount of information that was gathered. Yet instead of increasing anxiety, this informational surplus appeared to reduce it, as seems to happen when people gather information about their lives and worlds. So I hypothesized that there must be some kind of cognitive organization occurring outside awareness, perhaps the development of a neural map of the family that is implicitly coherent.
>
> (Landaiche, 2015, n.p.)

Over time, two questions kept recurring: (1) How does the study of one's system produce the differentiation of self that Bowen described even when the concepts of the theory are not clearly understood? And (2) How does the pursuit of one's own goals also benefit the group?

In regard to the second question, I believe we sometimes disparage self-interest as selfish and narcissistic. But I also see that those individuals who most successfully attend to their own particular interests also recognize their interdependence with others. Speaking for myself, I need others to do well in order to do well for myself. And the well-being of others is, I believe, also contingent on my doing well. Conversely, when I pursue self-interest while ignoring what matters to others, my degree of self-satisfaction will actually be greatly impoverished and compromised, often to the point of failure. At the most extreme, our sociopathic tendencies are based on an urgent need to believe that our omnipotence will yield satisfaction when in reality we need to work collaboratively with others to achieve what matters most to each of us individually.

I think this interdependent aspect of aspirational life bears directly on the interdependent aspects of leadership and followership in groups.

Leadership/followership

In regard to evaluating group leadership, Berne (1968b) took the position that:

> It matters little whether or not the group is "well run," or whether "that was a good meeting" according to some irrelevant standard; the only relevant criteria for judging *anything* that [the leader] does or does not do is whether the individual patients get well faster as a result.
>
> (p. 290)

Berne is advocating that we look at leadership behavior primarily in terms of its effect on desired or stated outcomes. Yet Bion (1959/1969) urges us to attend, as well, to the more emotional or less rational aspects of leadership, as when he cautioned, "A group structure in which one member is a god, either established or discredited, has a very limited usefulness" (p. 56).

From any number of these angles, leadership is arguably one of the most discussed and debated aspects of group life. It can be valorized and glamorized in the direction of Bion's "god, either established or discredited." Or it can be characterized more humbly as a form of service to the group's goals, which Berne saw as "the only relevant criteria" for evaluating leadership. And certainly—whether aggrandizing or humbling—leadership is often framed in terms of individual traits or behaviors.

As an alternative, I approach leadership from a group-as-a-whole or natural systems perspective, one that considers the group to be a living entity in its own right. From this perspective, leadership is an emergent property of the group, one inextricably linked with that of followership. In other words, there can be no leadership without followership, no directional movement of the group that does not involve the whole. And the way a group lives these reciprocal dynamics determines the quality of its leadership/followership, whether emerging in relation to certain tasks, particular anxieties, or specific challenges of learning. These give us measures of the group's overall functioning or maturity.

As Bion (1959/1969) wrote in his classic *Experiences in Groups*: I shall assume ... that unless a group actively disavows its leader it is, in fact, following him ... [That is, they are following] if ... the group shows no outward sign of repudiating the lead they are given (p. 58).

Bion is talking about the group's collective quality, the way passivity is itself a form of activity and the way directives from a designated leader, if actively opposed or simply not followed, point toward the actual leadership direction emerging from some other area of the group.

When leadership is assigned to a single member of the group, irrespective of the followers' roles in that process, then leadership is effectively a form of scapegoating. That is, someone in the group—perhaps the designated leader, perhaps a designated member—has been singled out, essentially to carry the burden or emotional pain for the rest of the group, with the psychotic belief being that, given someone to carry these things—to blame for these things—the rest of the group is now free. In fact, with a scapegoat in tow, ready to be thrown to the wild, the group has not only avoided the real problem, its resolution has been made even harder.

I am distinguishing here between the designated leader and actual leadership because I see the actual directionality offered to and followed by group members as emerging from anywhere in the group. The designated leader may not be providing leadership, but that is not necessarily an indicator of his or her failings as an individual. Rather, the group has decided, typically without explicit conversation, to follow another lead. Again, the group's decision is not necessarily a problem. There may be wisdom in the group that suggests moving in a direction other than that provided by the designated leader.

As discussed previously, Bion also wrote about the intensely compelling quality of a group's basic assumptions—to fight, flee, or engage in fantasies of salvation. These basic assumptions about group life do fit many of our histories of being either attacked from within our groups or attacked by outside groups. Moreover, fantasies of some savior-leader can carry enormous appeal when contrasted with the long, arduous haul of collective maturing. It can also be incredibly difficult to develop trust in a group that may be actually working together while making some serious, sometimes hurtful blunders along the way. As a result of these hesitations, Bion (1959/1969) wrote, "Leaders who neither fight nor run away are not easily understood" (p. 65). And on a wrier note, he suggested, "Should the therapist suspect that his high opinion of himself is shared by the group, he should ask himself if his leadership has begun to correspond with that demanded by the basic assumption of the group" (pp. 73–74).

The primal forces in our groups can be difficult to weather and challenging to extricate ourselves from without being "killed" by the group that is not yet ready for such a productive developmental step.

This powerful, primal intensity is related to what Bowen (1978/1994) wrote about in terms of the tension between separateness and togetherness in families and other human groups. It can be quite difficult to define an authentic, individual self in the face of the family's or group's intense anxieties and hostilities. And it is even more impossible for the group to develop maturity when its members are unable to define themselves as separate from the group as a whole, as individuated from what Bowen called the "undifferentiated family ego mass" (p. 159).

This capacity for separating as an individual from the group's compelling togetherness—while still remaining in contact, without cutting off—bears directly on the functioning of leadership and followership in groups that learn. According to Bowen (1988), a family or group leader is one

> with the courage to define self, who is as invested in the welfare of the family [or group] as in self, who is neither angry nor dogmatic, whose energy goes to changing self rather than telling others what they should do, who can know and respect the multiple opinions of others, who can modify self in response to the strength of the group, and who is not influenced by the irresponsible opinions of others ... A family [or group] leader is beyond the popular notion of *power*. A responsible family [or group] leader automatically generates mature leadership qualities in other family [or group] members who are to follow.
>
> (pp. 342–343)

Such a leader could be any member of a family or group with the strength and courage to manage reactivity and take a principled stand. As my first group teacher, Nick Hanna, used to say, "Good group leadership is good group membership" (personal communication, circa 1989).

In the words of organizational consultants Leslie Ann Fox and Katharine Gratwick Baker (2009), leadership is "a reciprocal process ... a relationship rather than a set of individual characteristics" (p. 16). This suggests that we look at leadership and followership as two dimensions of a single, necessarily mutual dynamic of groups, one in which a lead or direction is simultaneously given and followed. There is a necessary interdependency and flow. In addition, there is implicitly, if not explicitly, a contractual nature to this reciprocal relationship, an agreement (even if never spoken) to engage in particular working conditions and to adopt particular roles and rules.

In Chapter 5, I wrote about the process of contracting for groups, which Berne (1963, 1966) covered in his books on organizational and group processes. This notion of contracting has also figured centrally in the practice of transactional analysis and has been the subject of varied applications and intense discussions (see for example, Blacklidge, 1979; Drye, 1980; English, 1975b; Gellert & Wilson, 1978; Gibson,1974; Haimowitz, 1973; Lee, 1997; Loomis, 1982; Maquet, 2012; McGrath, 1994; Moiso, 1976; Mothersole, 1996; Roberts, 1984; Stummer, 2002; Terlato, 2017; Vanwynsberghe, 1998; Weiss & Weiss, 1998; White, 1999, 2001; Wright, 1977).

In regard to the leadership/followership dynamic, I see there being three salient manifestations: one related to particular tasks, another related to a group's anxieties or felt intensities, and a third related to the processes of effective learning. Each of these involves different forms of contracting, which may be conducted in Berne's (1966) bilateral spirit—achieved multilaterally in groups—or agreed to outside of conscious, informed consent. Thus the three constructs I work with are:

- **Task Leadership/Followership.** "Task" roles refer to formal designations as well as to administrative and practical divisions of labor, which may vary from group to group. For example, the timekeeper is the leader dependent on the cooperation of those following, as is the person who sets the structure, collects the money, and so on. Contracting in regard to tasks—sometimes shared among members of a group—is typically more explicit and straightforward, although not always.
- **Anxiety Leadership/Followership.** "Anxiety" roles refer to interactions and decision-making motivated by threats and expediency, more instinctive than thoughtful, and developmentally less mature albeit still important in situations that require automatic coordination of the group effort. For example, an anxious group may trigger directionality in one individual who is simultaneously followed by the rest of the group to fight, flee from danger, or suppress the emergence of complex thought or individual expression. These roles are instinctual and can be projective, scapegoated, scripted, protocol level, acutely and chronically activated, and/or at times psychotic, to name a few possibilities. Contracting in regard to anxiety roles is almost always implicit, often outside awareness, and sometimes coerced on both sides.
- **Learning Leadership/Followership.** "Learning" roles refer to interactions that promote or facilitate maturing and development of the individuals in the group and of the group itself. These roles show considerable flexibility as individuals move easily into and out of leading and following based on the changing nature of the learning needs and the evolving resources to be found in the group. There is balance in attending to the needs of the group and of the individuals in it. Contracting is fluid, responsive to the process, and if not fully conscious, then easily amenable to being talked about.

Depending on circumstances and on the group itself, these three aspects express, respectively, the group's practical needs, its emotional process, and its capacity for maturing, even as these aspects serve to divide the group's labor.

I have been thinking lately about the degree to which these leadership/followership roles emerge spontaneously, as certainly happens in anxious situations. However, I think we can also see the instinctive emergence of leadership/followership in situations in which learning emerges in a free and creative way. I have been contrasting these more naturally emergent forms with those I see as more intentionally human made: structures, like tools, that can be practical or task oriented (supporting the status quo); structures of leadership/followership that may represent institutionalized forms of anxiety (repeated without thought); and structures of leadership/followership that are employed intentionally, perhaps with discipline, as means of facilitating learning and based on prior effectiveness.

For now, I suggest that group manifestations of leadership/followership appear to differ based on the degree of separateness and togetherness in the group itself, which is to say, based on the group's degree of chronic anxiety or its perception of the intense threat that Berne (1963) believed to be so important in understanding group dynamics. And although certain individuals may be conscripted into particular leadership/followership roles based on individual proclivities or vulnerabilities, it remains important, I believe, to see these roles as existing for the group and as representing the group's maturational functioning. As Bowen (1978/1994) wrote:

> A family with a psychotic family member is a functionally helpless organism, without a leader, and with a high level of overt anxiety. It has dealt helplessly and noneffectively with life, it has become dependent on outside experts for advice and guidance.
> (p. 79)

In contrast, Bowen (1978/1994) described a family that begins to pull itself together and begins to enlist its individual members to tackle the various problems facing the family:

> When the family is able to be a contained unit, and there is a family leader with motivation to define the problem and to back his [or her] own convictions in taking appropriate action, the family can change from a directionless, anxiety-ridden, floundering unit, to a more resourceful organism with a problem to be solved.
> (p. 85)

I have certainly seen anxiety leadership/followership in my own family, as I described earlier, including in my functioning as the oldest of nine children in which my "leadership" style was framed as follows: (1) look like the one in charge, (2) watch out for everyone else's welfare, (3) keep a low profile to make sure you don't get "killed," and (4) keep everyone in order and well-behaved.

In some contexts, this might sound great, but I found it damaging in practice. After all, it conveys: (1) leadership is a role, a costume, not a set of qualities one grows into; (2) I don't exist as an individual; (3) the group members (the "follower siblings") have no minds of their own; and (4) we avoid the messes and actual problem-solving that our family needs for growing and learning.

Bowen (1988) himself wrote about recognizing his own role in the life of his organizational groups: "I played a part in any problem or symptom that developed in the staff, and ... the disharmony would ... be corrected when I had modified my part in the creation of the problem" (p. 343).

In my family there were no explicit messages about following because the followers did not really exist as individuals, just as I did not. It took me many years to learn that there are times when following someone else's lead is the best way to achieve my own goals. I just could not do it alone. I now think of both leading and following as potentially mature, effective choices, just as I am continually alert to their potentially less-mature manifestations. In this regard, I particularly appreciate organizational consultant Erik Thompson's (2011) teachings about the importance of finding courage when learning to lead—and, I contend, learning to follow.

Therefore, the facilitative practices I will describe are not meant for the group's designated leader alone. They must be cultivated and incorporated over time into the learning group's culture. The designated leader may propose certain principles and practices for the group to follow, but that does not necessarily mean the group will choose to follow that particular lead. In such instances, how does the committed and courageous designated leader live those principles and practices while accepting that the group chooses to follow another lead? How would a committed and courageous designated member do likewise? Instead of fighting with the group, how do we each fight for our own principles and values, our own integrity?

Actual leadership of this sort can emerge from anyone in the group, although this does not mean that the group will follow. But I believe it is the only foundation on which actual group learning and maturing can be built.

Defining a self in one's system

The interplay between followership and leadership speaks to one paradoxical aspect of living in groups. For alongside the collective dimension to human life, there persists an essential individualist factor. We can only truly be part of the group to the extent that we find and define our individuality, just as we each can only become ourselves by knowing and living our places within our larger human systems.

One of the things I find most useful about Bowen's theory is that it is meant to be lived, not just thought about. And I think he provided us with some clear instructions for what to do and how to live it. He outlined practices (Bowen, 1978/1994) I find useful for facilitating learning for myself and for others in my groups. Interestingly, these practices also outline an approach to researching and studying ourselves in collective life.

In my own words, I would state these practices as follows:

1. Making contact with the emotional system
2. Observing oneself in the emotional system
3. Managing oneself in the emotional system
4. Establishing one-to-one contact with other individuals in the emotional system

I see these as stages in a process. That is, before I can do anything else, I have to make contact with the system, with the group. That contact may not look (or feel) very pretty, but it does not have to. It just has to be contact, which I would counterpose to cutoff from the system, that is, disconnecting from the reality of our social being. This applies whether we are making contact with a family, a community, an organization, or any other kind of human group. I would elaborate this first stage also in terms of adopting receptivity, of opening myself to the contact and its vital systemic information.

Once I have made contact, there may come a day when I can also observe myself in the system—observe myself in my own bodily system and in the larger body of my family, community, or organization. Something happens when I observe, even if I can bring myself to do nothing else that is productive, even when I feel or behave like the same old mess I have always been. I therefore elaborate this stage to include the concept of suffering (as discussed in Chapter 1) as a willingness to bear the facts of the world and to remain doggedly committed to seeking even if we are not happy with what we find. This speaks to developing a greater bodily capacity for life's challenges as well as its more intense pleasures.

On better days—while in contact and observing—I find myself also able to manage myself in my system, to make choices not to engage in behaviors that do not serve my own goals or the well-being of the group, maybe once in a while do more than just *not do* but actually *do something* productive. If nothing else, I will settle simply for containing myself and my reactive actions. This practice of managing myself also permits the integration of experience that I find so necessary for pursuing what matters most to me.

Finally, there are even fewer days—although oh so glorious—when I am in contact, observant, and managing my reactive actions and can also speak of myself to others in my system. At those time, I can actually be interested in them as individuals, not just see them in terms of their familiar functions in the system. This stage of action includes making statements for and about myself as well as listening to and for the separateness of others.

Although I often use these four practices in the context of me in my family, in my experience they can be applied in any system. And I especially find they can be cultivated in groups.

Facilitative practices

The facilitative practices of learning continually and experientially, of holding space for everyone to speak and think freely for herself or himself, of listening to the bodily effect of the conversation's content, and of sharing hypotheses while researching further—all of these work in concurrent synergy. Yet I find it helpful to talk about each practice separately, to highlight their distinctive aspects.

I also want to contextualize these facilitative practices as part of the multigenerational transmission process, a form of reproduction that is one of the three essential characteristics—along with metabolization and responsiveness—of any living system. As I said in a presentation I gave in Pittsburgh for the 2105 international conference on Bowen theory:

> Reproducing life *requires* the transfer of information—everything from DNA to instruction sets for sheltering, feeding, child-rearing, and warring. And in any species for which the transferred information and relevant behavioral sequences cannot simply be activated automatically via genetic or epigenetic expression, some form of instructing is key to preparing the next generation.
>
> ... And since that instructing and training must meet acceptable minimum standards for survival over time—not only of the immediate young but of succeeding generations—such training is an activity of urgent interest to the community, sometimes even an activity of the *whole* community.
>
> ... To what extent can we influence that succession? When does that influence serve our principles more than our reactivity?
>
> (Landaiche, 2015)

In that spirit, I want to identify the structures that facilitate learning for individuals in the group (and for the group as a whole) as well as the possibilities for transmitting those structures, as knowledge, to the next generation. Such intentional transmission is based on the hunch that the structures will, at minimum, sustain the learning capacity of the group if not increase it ever so slightly over time, within realistic limits.

These facilitative, transgenerational structures are employed by learning communities that follow a set of principles and use particular methods or practices, all of which, in turn, are passed to the next generation. Kuhn (1962/2012b) describes this in relation to the transmission of scientific paradigms, using the gender-normative terms of his particular cultural paradigm:

> The study of paradigms ... is what mainly prepares the student for membership in the particular scientific community with which he

> will later practice. Because he there joins men who learned the bases of their field from the same concrete models, his subsequent practice will seldom evoke overt disagreement over fundamentals. Men whose research is based on shared paradigms are committed to the same rules and standards for scientific practice. That commitment and the apparent consensus it produces are prerequisites for normal science, i.e., for the genesis and continuation of a particular research tradition.
>
> (p. 11)

I find it helpful to think of these transgenerational structures in developmental terms. That is, they represent the best thinking of the elder generation, much of which they also likely inherited from prior generations. When passed along to the newest generation, that received transmission of wisdom offers a place from which to begin a learning process, without having to start from scratch. I see the transmission of principles and practices as facilitative in that sense, as a structure within which to begin a potentially more complex learning process. I think Kuhn acknowledged this helpful function of what he called a *paradigm* even as he also cautioned us to be aware that sometimes what is transmitted may no longer represent actual thinking or researching but may simply have become dogma. The problem is not that the paradigm or facilitative principles and practices are not perfect or eternal. The problem occurs when they remain closed to future scrutiny and experiential evidence of a need for a newer way of understanding and of responding to life.

So, I use the term *principles* to refer to general guidelines and values from which particular behaviors or activities are derived. *Methods* or *practices* are the particular behaviors or activities that have been shown to support the principles of the learning community and to facilitate the community's objectives. Their implementation can be modeled by any and all group members, whatever their formal roles. It is up to each individual in the group to decide whether to adopt and commit to them—with authenticity and integrity, not simply from conformity.

Learning continually

In relation to the community or systemic context for learning, I wrote earlier about appreciating the words of radical educator and philosopher Paulo Freire (1970/2000) concerning the teacher being "no longer merely the-one-who-teaches, but one who is himself [or herself] taught in dialogue with the students" (p. 67).

This fits my observation over time: I have learned most from teachers and mentors who have remained engaged in their own learning processes, just as pursuing my own learning seems correlated with greater effectiveness when facilitating the learning of others. Learning in groups seems to be one outcome of the collective, interactive dialogue among facilitators, students, and clients that Freire referenced.

This means that to be effective in my professional facilitative role, I need "to learn something new every week" (Berne, 1966, p. 61). Specifically, I need to keep learning generally about how humans learn (and have trouble learning) as well as to keep learning about my own learning process and challenges. And because, as I have written, I see group life as unavoidable for human beings, some of what I need to learn can only occur within a group, with colleagues, family, my work organization, and so on.

This first principle and practice of "learning continually" is based on my observation that learning—my own and others'—is affected (for better or worse) by the neurophysiological states of significant others, by their levels of chronic anxiety as well as by their capacities for managing reactivity and defining a self with some integrity.

To be in the presence of this fruitful physiological learning or integrating process with significant others potentiates this same neurophysiological process in ourselves as individuals and groups. Conversely, to be in the presence of the nonlearning neurophysiological state (especially that of the facilitator or designated leader) can also inhibit this learning process. Nonlearning, as a form of chronic anxiety, can be as contagious as learning. Yet learning's infectiousness appears to lower reactivity and to increase the capacity for refusing impulsive action.

In our professional facilitation roles, we may be characterized as engaged researchers, actively involved in gathering experiential data and in theorizing human life. At times, we may be containing that information without speaking until the data have sufficiently organized themselves, which is to say, eventually speaking from integration rather than from emotional process or blurting out. Most importantly, we attempt *not* to bring into our groups the deadness that comes from having abandoned or despaired of the process of learning anew.

Learning experientially

Experiential learning is typically contrasted with learning from books or didactic content. And although my own most important learning—about human growth and development—has certainly been experiential, the irony is not lost on me that my most important learning about experiential learning has come from educational theorist David Kolb's (1984) book by that same name. I find his analysis of the concept to be exceptional. He also showed the connections among many different thinkers, in effect offering a more integrated foundation for figuring out how to structure learning experiences that are intentionally experiential as a means of deepening their learning impact.

Kolb (1984) described a process whereby experiences—whether pursued intentionally or by chance—are later reflected on via a movement back and forth between action and study, a "dialectic tension and conflict between immediate, concrete experience and analytic detachment" (p. 9). In teaching Bowen theory, for example, I have found that there is a necessary interplay between learning theory and learning from direct research experience about one's own family system.

I think there is also an important dimension of this learning that Berne (1949/1977d) articulated with his conceptualization of intuition:

> Intuition is knowledge based on experience and acquired through sensory contact with the subject [the other], without the "intuiter" being able to formulate to himself or others exactly how he came to his conclusions ... It is knowledge based on experience and acquired by means of preverbal unconscious or preconscious functions ... The individual can know something without knowing how he knows it ... Not only is the individual unaware of how he knows something; he may not even know what it is that he knows, but behaves or reacts in a specific way as if ... his actions or reactions were based on something that he knew.
>
> (pp. 4–5)

In effect, our experiences of being in groups will exceed our conscious capacities. And yet, as Berne described, we still have ways of formulating those experiences, ways of adding them to our stores of knowledge. Bion (1962/1977c) also wrote about learning from experience in his similarly titled book.

Everything, of course, is a form of experience. Think about what it is like to be forced to listen to dry, technical lecture material and then to spit it back out exactly as delivered under pain of punishment in the form of a bad grade. Although some information is useful to know by rote, one learns little in that educational context about one's own learning experience or how to facilitate such experiences for others. This book, in fact, offers an experience that is largely didactic. Yet I hope there is in my words enough vitality to actually affect your experience in relation to the ideas, enough vibrancy for you to carry into your next group encounter. That is my aspiration.

Thinking and speaking freely

The ability to think freely, without preconceptions, and then to speak authentically of what we each think are ideals not easily achieved in a group. In that context, the welter of experiential data may be such that we cannot find even the sense of an intuitive mind, much less a conscious one. And then when we do discover what we think, it is not always easy to bring that to the group, as poet and writer, Czeslaw Milosz (1969/1975) acknowledged:

> Each of us is so ashamed of his own helplessness and ignorance that he considers it appropriate to communicate only what he thinks others will understand. There are, however, times when somehow we slowly divest ourselves of that shame and begin to speak openly about all the things we do not understand.
>
> (pp. 3–4)

In effect, we may sometimes find the courage to move slowly past that shame or social pain in order to speak with openness. And in speaking of the "things we do not understand," we may be giving voice to the intuition about which Berne wrote.

Another related practice, taken from the psychoanalytic tradition and endorsed by Bion, is *free association*. Psychoanalyst Christopher Bollas (2009) wrote about this as a method "designed to reveal a 'train of thought'" (p. 6). He also suggested that we "redefine free association as *free talking*, as nothing more than talking about what is on the mind, moving from one topic to another in a freely moving sequence that does not follow an agenda" (p. 8) but rather one that "reveals a line of thought ... linked by some hidden logic that connects seemingly disconnected ideas" (p. 6). He added that sometimes "the most valued material is the apparently 'irrelevant'" (p. 8).

Some of what may be most useful for the individual and the group may be what feels most risky to voice aloud. Indeed, in an anxious, nonlearning group, the injunctions against such freedom may be powerful and carry equally powerful punishments, such as shaming. Yet can we find our courage?

For in sharing openly what we have learned or are in the process of learning—even what we do not yet understand—it seems that the group's dialogue becomes more creative and productive. We break out of the older, no longer workable paradigms when we allow ourselves to think, to freely associate, and to speak more openly, again assuming we are speaking with authenticity rather than from unmanageable reactivity.

I find that we can feel the difference between anxious blabbering and the emergence of what I refer to as a seemingly nonsensical poetic utterance. The former contributes to the all-too-familiar group atmosphere of oppression and suffocation, whereas the latter cultivates a group dynamic that is fueled by principled aspiration. Even when our words do not make logical sense, speaking freely and authentically appears to lead us to have greater confidence in the brain's natural ability, as well as the group's, to form connections, not all of which will be directly available to consciousness. It also develops recognition of the importance of speaking freely for human maturation, self-awareness, clearer communication, and creativity.

We are a species that makes use of language with the sometimes salubrious outcome of giving life to our groups.

Speaking for self

At times, when I am the designated leader of a group and given plenty of airtime, I may be so infected by the group's anxiety that, in effect, I am speaking for the group, not for myself. I put the group's anxiety into action and effectively do not exist as an individual. When the group's anxieties are that high, I am less aware when another group member is speaking for the group, just as I am less sensitive to the expressions of authenticity that may be intermixed.

As a member of my groups, I can avoid some nonauthentic speech when I ask myself if what I am voicing represents my experience, my sense of what is true. I can avoid nonauthentic speech by striving to make "I" statements, by speaking for myself and of my own experiences, by honoring my own awareness, feelings, and ideas. I can take a more humble stand if I do not presume to speak for others or attempt to force them to see things my way.

When I am the designated leader, I may reflect back what I hear a designated group member saying. I may strive to reflect back what I think is the more authentic statement that individual is making about herself or himself. Or I may, in listening to one group member, offer my sense of what is being expressed for the group, an expression of what we may be struggling with as a group. I may even be able to make use of my own reactivity to reflect on this stage of our existence as a group.

When all of us as group members are able to speak for ourselves, we make greater room for the kinds of differences that will nurture the group's future life.

Holding space for everyone

Not all group members will use the same means of communication. Some will talk, of course. Some may signal their participation nonverbally with facial expressions, particular kinds of breathing, or bodily gestures. Some may use words differently. Some may feel more comfortable dominating the group than others. Some may hide themselves in a kind of noisome silence, whereas others may chatter on while revealing little.

Given these differences, it can be difficult to hold space in a group for everyone to contribute. And yet attending to the different individuals in the group—usually by making room for everyone to speak—can be a powerful means of acknowledging other individualities and differences while also welcoming authentic free speech and free association.

As the designated leader, I may do this by asking open-ended questions that will help group members elaborate experiences and expand their awareness. I may, as indicated earlier, reflect back what I hear or understand as a way of verifying the accuracy of my reception. I may also take what individuals are reporting and link that to my hypothesis about what is going on at an organizational, community, or group level, thereby supporting a broader systems perspective for the group. When group members become stuck advising, arguing with, or otherwise abandoning self, I will look for a way to guide each individual back to self-expression and more exploratory behaviors and speech. I may also invite participation from quieter group members, taking care to notice when their quiet has also been expressive of something I might then want to put into words for the larger group.

Listening to the content and its bodily effect

Groups of humans yakking away are rarely short of content. In fact, the content in a group may be overwhelming in quantity and quality, particularly when the group is working productively. So although I believe it is important to attend to the content in a group discussion, I also do not want to lose track of what may be going on behind it or the way a particular content is being used to achieve something in the group. I make an effort also to listen to the effect of that content on my body as well as on the body of the group as a whole. In the literature on groups, this is sometimes described as attending to the content and process of a group. It is important to make use of the content's bodily effect as a means of making provisional interpretations of meaning.

Sharing hypotheses

When listening to an individual or group, we may sometimes offer our thoughts—our interpretations—of what we are hearing, both content wise and in terms of our bodily experiences of the group and its members. If our interpretations are offered provisionally rather than dogmatically, we are, in effect, posing a hypothesis that others can reflect on, measure how it might fit with their own experiences, and even disagree with in a manner that is productive for the work of the group.

When we share our hypotheses as they emerge and evolve, this constitutes a form of verbal integration or interpretation of the group's learning process and discoveries. It is important to then observe the effects of such spoken hypothesizing or interpreting to see if the words help to settle or downregulate the group's reactivity, to see if the language helps to integrate experience before that which is nonintegrated becomes too overwhelming for the group and thus fuel for destructiveness.

Sharing hypotheses about our learning process also helps to bring to greater awareness the ideas that are operating as a paradigm. It makes those ideas more available to scrutiny as well as to their intentional transmission to the next generation.

Researching further—studying ourselves as ongoing processes in collective life

I do not need to say much more about the practice of researching further because that has been the theme of this entire book. However, I think it important to note the relief that can come from not having to know, from being able to face the essentially infinite quality of group life with an attitude of openness to the ongoing process, to accept with humility and a sense of hope that what in the moment may seem so disorganized and disorganizing may eventually find some order. We just may get through this together.

Berne (1972) addressed this process of researching directly when we wrote, "The first duty of a group therapist ... is to observe every movement of every muscle of every patient during every second of the group meeting" (p. 315). And yet he also let us know that, in writing *The Structure and Dynamics of Organizations and Groups* (Berne, 1963), he drew on a wellspring of experience:

> This study is based on a schedule of leading, observing and participating in groups over a period of 19 years, as well as teaching and supervising group therapists and acting as consultant to leaders of ailing groups and organizations of various kinds. This included experience with about 5000 shifting situations of different types with the Army, the Navy, the Veterans Administration, the California State Hospital and Correctional systems, municipal, county and private community service agencies, the University of California and Stanford University. The responsibilities and pleasures of everyday living have also offered learning opportunities in private psychiatric practice, politics, athletics, religion, education, science, the classroom, the courtroom, the theatre, at the buffet and around the campfire.
>
> (p. vii)

These kinds of basic research are important, especially when the community's learning is centered on human functioning. This means studying ourselves as ongoing processes—our histories, collective behaviors, reactions, aspirations, areas of vulnerability, and skills. This would be partly a way to learn generally about other human beings and partly a way to distinguish or differentiate oneself from others.

As always, we can then bring our necessarily incomplete research findings to the group and, in pooling such resources, pave the way for further study and learning.

Finally, in addition to these potentially facilitating behaviors, other constructive tools for the group might include: a boundary between the group and the larger world that might involve an agreement for confidentiality, a commonly agreed on focus for work, an agreed on time and space structure, an ethos of not harming, permission to speak freely in turn as well as permission to remain silent, suggested experiential learning exercises, and finally, the capture of group learning—in the form of writing or other forms of symbolization—for the purpose of transgenerational transmission.

10

CLOSING REFLECTIONS

For many years now, I have been interested in learning as my primary purpose, activity, and aspiration, my reason for being, in one sense my fate, yet something I can also well avoid. I don't have to live it. At the same time, my learning has also been the function of an inheritance, my good fortune to have had effective teachers, mentors, and forebears as well as conducive groups. My learning has been affected profoundly and productively by others and would not have been possible without them.

My process appears also to have affected others, albeit sometimes in the direction of nonlearning, as when I cut off, overfunction, triangle (i.e., engage with others primarily to discharge anxiety), enact third-degree games, or am otherwise not my genuine self.

In one sense, my concern with the next generation seems automatic, built into my biological being, not really something I do consciously to fulfill an ideal. Although it undoubtedly fulfills my being and genetic heritage, certain of my automatic impulses to parent, teach, or mentor are not always thoughtful, as when my more anxious, fatherly impulses toward my daughter, for example, sometimes have to be curbed in order to give her the room she needs to grow toward the person my less anxious self would actually want her to become.

When I think of life as more than just a single instance but rather a transgenerational phenomenon, it has become apparent that one aspect of human maturing involves managing tensions, over time, between the individual and the group.

We seem also tasked with managing the tensions between older ideas and those still emerging, which Kuhn wrote about in terms of paradigm shifts. As philosopher James Marcum (2015) wrote, "The older paradigm is like a fossil; it reminds the community of its history but it no longer represents its future" (p. 67).

This idea of a fossil that no longer represents the community or family's future touches a deep sense of loss that haunts me still. It is a puzzling emotional experience because one might say that writing here was essentially an act of creation, a gain. Yet there is something lost for me in giving up the old models, the old hypotheses that, even if unworkable, still had the feeling of familiarity, of family, one might say.

Yet my hope is for myself and my groups to function at capacity, to the best of our human abilities. As I have explored what I can do to facilitate such learning, it has challenged me to rethink and rearticulate what I mean by "learning" and, in some cases, to dismantle some of my most cherished beliefs. More questions keep erupting to the surface. And the work continues for as long as we persist as a human species, with its attendant losses and gains.

I characterize this as glancing back and straining forward from the unbearable present.

Human intelligence or cleverness?

The concept of learning has been broadened in all fields to include more than just what small children do to grow up or students in school do to master material. It is now viewed as something that scientists and professionals need to continue for the whole of their lives as well as a process that can potentiate learning in organizations and groups of all kinds for the ongoing needs of adaptation to change. Yet although I have some hope for the potential of human learning communities to serve human life, I also want to temper that optimism by distinguishing between human cleverness and human intelligence or wisdom.

That is, humans have proven themselves absolutely ingenious when it comes to making tools and all manner of things. But just because we can make something, just because we had to use our brains to do so, just because one innovation leads so seamlessly, even compellingly to the next, I do not believe that means we have used our intelligence or our capacities to make decisions based on more complex assessments of the situational data.

I propose that humans are clever enough to make highly sophisticated tools that do not represent actual intelligence or wisdom but that instead perpetuate the chronic anxiety that bore them forth in the first place. For example, the university where I work is a leading center of technology and innovation. Yet our university president gave a talk at the start of one academic year in which he described how the greatest global challenges we face today—overpopulation, environmental degradation, and so forth—are the result of the greatest technological achievements of the prior century. As we march determinedly forward with what we call progress, we seem to be cunningly creating some of our most precarious life dilemmas.

In Yuval Harari's (2011/2015) book *Sapiens: A Brief History of Humankind*, he called the agricultural revolution "history's biggest fraud" (p. 79). His position, I learned later, is a riff on a similar position taken by professor of geography Jared Diamond (1987) in an earlier article entitled "The Worst Mistake in the History of the Human Race." Basically, all of the alleged benefits of agriculture do not seem to stack up well against the costs in labor, poorer nutrition, overpopulation, greater exposure to diseases, increasing income disparities, and continual investments in an infrastructure that must be frantically maintained primarily because it now seems too costly, and so too late, to turn back. Harari (2011/2015) concluded:

> These plants [wheat, rice, and potatoes] domesticated *Homo sapiens*, rather than vice versa ... According to the basic evolutionary criteria of survival and reproduction, wheat has become one of the most successful plants in the history of the earth ... [And] wheat did it by manipulating *Homo sapiens* to its advantage.
>
> (p. 80)

Harari (2011/2015) then situated the industrial and subsequent technological revolutions in what he saw as direct outgrowths of the same human processes operating for agriculture:

> Nowadays I can dash off an email, send it halfway around the globe, and (if my addressee is online) receive a reply a minute later. I've saved all that trouble [writing, addressing and stamping an envelope, taking it to the mailbox] and [saved all that] time [days, weeks, or even months to get a reply], but do I live a more relaxed life?
>
> Sadly not ... Today I receive dozens of emails each day, all from people who expect a prompt reply. We thought we were saving time; instead we revved up the treadmill of life to ten times its former speed and made our days more anxious and agitated ...
>
> Humanity's search for an easier life released immense forces of change that transformed the world in ways nobody envisioned or wanted. Nobody plotted the Agricultural Revolution or sought human dependence on cereal cultivation. A series of trivial decisions aimed mostly at filling a few stomachs and gaining a little security had the cumulative effect of forcing ancient foragers to spend their days carrying water buckets under a scorching sun.
>
> (p. 88)

The point here is not, I think, to romanticize hunter-gatherer lifestyles but rather to question the assumption that our technologies necessarily represent improvements.

Rather, there seems to be evidence that human cognitive abilities can be put to work for reactive, emotional process ends, sometimes responsive to actual situations or threats and other times responsive to anxieties that operate chronically, across generations, in the absence of any actual or immediate danger. Can we determine what tools represent a response to actual conditions, thus perpetuating life, reducing harm, and benefiting the next generation and perhaps even the larger, ongoing world? Conversely, what tools contribute to more chronic anxiety, to maladaptive responsiveness?

The atomic bomb would seem the literal exemplar of the arms race for escalatingly clever and destructive toolmaking. Perhaps alongside the more obviously devastating aspects of that particular technology, we might point to the collateral gains—such as nuclear powered plants and submarines—although even those putative benefits remain to be weighed against their corresponding costs and contribution to further unsustainable escalations.

I have been using the term *tools* to reference all manner of human-made items, which might include shelters, roadways, agriculture, and so on. But is there a difference between human toolmaking and human symbolizing in language, art, music, dance, poetry, and so forth? Is there a difference between human-made objects and human-made meanings?

For example, what are we to make of human writing and, in particular, mass publication? On the one hand, such publication seems essential to multigenerational transmission, including the work of the scientific community (in its nonchronically anxious mode). Yet we would also assume that publication contributes as well to the transmission of chronic anxiety, even mass hysteria. We can certainly see this in certain uses of social media. But does publication *necessarily* contribute to learning, to the group's maturing? I know that the published works of others have been important for my learning, as have certain works of art, along with the long tradition of meaning making so central to the humanities. Yet if I look at the full range of human publications and media, how would I distinguish their learning and nonlearning aspects?

These are the kind of evaluative questions that I propose are important when assessing human learning processes and outcomes. Evidence of mental capacities and brainy activities does not mean that learning is taking place; sometimes it is the obverse.

And if our species operates with a degree of chronic anxiety that harms us and possibly leads to our extinction, is there anything I as an individual can realistically do to help shift our functioning? I do believe there is at least a different way to show up amidst the emotional process, one that may have more of a quality of mercy than of substantively redirecting our fate. This would bear on my ability to remain a viable member of my human groups.

Progress and mortality

Now that I have thrown cold water in the face of human cleverness, my second caveat pertains to the notion of human progress, which as Kuhn (1962/2012b) wrote, is "a perquisite reserved almost exclusively for the activities we call science ... [and not a requirement of,] ... say, art, political theory, or philosophy" (p. 159). Indeed, the lack of any notion or expectation of "progress" has been part of what has drawn me to the arts and humanities as evolving expressions of signification for human life.

Yet Kuhn also questioned the notion of science as representing progress, as moving toward something better or truer—even with its history of paradigm changes. As Ian Hacking (2012), philosopher of science, wrote in his introduction to the fiftieth anniversary edition of Kuhn's classic, "[Kuhn] ends with the disconcerting thought that progress in science is not a simple line leading to *the* truth. It is more progress *away* from less adequate conceptions of, and interactions with, the world" (p. xi).

In Kuhn's (1962/2012b) own words, "We may ... have to relinquish the notion, explicit or implicit, that changes of paradigm carry scientists and those who learn from them closer and closer to the truth" (p. 169).

Yet learning typically implies a developmental progression, a movement to increased integration and greater capacity. Might there also be learning required by the individual and the system that leaves the system or individual essentially at the status quo, treading water just enough? Is it possible to make a distinction between learning that maintains the status quo and learning that represents evolutionary movement?

Like Kuhn, I do not want to valorize "progress" as anything more than progressive adaptation to changing conditions. I want to distinguish between notions of teleological or goal-oriented progress and the progression across time that simply permits ongoing life, barring too large a shift in the environment that, for certain species, would press fatally beyond the limits of that life form's "ongoingness."

In that sense, mortality also poses a problem for the ideal of learning as progress, as our somehow moving closer and closer to defeating death. If nonlearning—the inability to integrate and adapt—puts a species at risk of extinction, what are we to make of the likelihood that every form of life will, at some point, have its day and see its demise? For although bacteria have demonstrated for millennia, perhaps more than any life form, some pretty serious learning in terms of how to survive wildly varying conditions, I still will not claim that bacteria are immortal.

So, if I think of a lifespan as existing for the individual as well as for the group or species, perhaps it is most realistic to see aspirational movement simply as the fulfillment of the genome across its achievable span of unfolding life. We achieve what is embedded within our biological, mortal bodies.

Hoping within reason

If, as I have proposed, we conceptualize learning as the neurophysiological integration of the data of experience—an integration that allows for the living organism's adjustment to reality—such learning to work within groups seems to be about working for the continuity of human life, for our existence as a social species that, like every form of life, has figured out how to reproduce itself in a manner that can live in the world today and the one most likely to come next. A group that learns is one that lives over time within a tolerable range of changing conditions. A group that cannot learn will likely perish more quickly.

Unintegrated (underhypothesized) experience, for human beings, can become so intense that its bodily violence leads to violent action, thus to harm, trauma, and destructiveness. Yet there can be, in the violent act—along with the impulse to eradicate—also a mobilization of energy necessary to care for the self-in-becoming, both to titrate the rate of incoming information and to direct it toward sturdier, preferable growth. I think we can see this operating for the individual, and I wonder if we can interpret systemic or group-level violence as also expressing a mobilization of aspiration, one that is protective, directional, and motivating even as it may sometimes tragically misfire and deliver its opposite, life-destroying outcome. In other words, in a family's or community's tragic history, can we discern the wisdom inherent in its protective impulses and detect the particular path the family or community might have taken had it only known how to manage the extremes of its emotional process?

Moreover, even if the geological record is replete with evidence of uncountable species' extinctions—suggesting that ours will also one day cease to exist in any recognizable form—still it seems inbuilt to our nature to strive for life's continuity into that uncertain future, much as we each, individually, may struggle mightily to extend our own individual lives in spite of death's eventual certainty (albeit a certainty with meaning that is separate from what some of us believe to be the existence of an everlasting afterlife). Mortality still seems to deserve its own respect and certainly can offer its own reward.

For me the question of whether groups can learn was also motivated, in good part, by my interest in the reactive, emotional processes of human societies. Can we as a species learn how to keep living, especially in terms of how to avoid going extinct?

To answer these larger questions, I have suggested the ongoing study of ourselves in collective life. Toward that end, Kuhn's (1969/2012a) words seem apropos here:

> Having opened this postscript by emphasizing the need to study the community structure of science, I shall close by underscoring the need for similar and, above all, for comparative study of the corresponding communities in other fields. How does one elect and how is one elected to membership in a particular community, scientific or not? What is the process and what are the stages of socialization to the group? What does the group collectively see as its goals; what deviations, individual or collective, will it tolerate; and how does it control the impermissible aberration?
>
> (p. 208)

Toward a more capacious theory and practice of group work: Berne, Bion, and Bowen

I am fortunate to have inherited the ideas of three generative forebears: Eric Berne, Wilfred Bion, and Murray Bowen. Each of their theoretical frameworks and methods of practice was distinctive, some might even say mutually exclusive. Perhaps certain aspects have even become fossilized, in Marcum's conception, certainly at times taught by rote. Yet to the extent that these theories have remained alive and evolving, I experience them as speaking to different facets of the same larger phenomenon of human life. Moreover, each of their efforts was toward figuring out how to resolve the troubles that have plagued human life for millennia, especially in our grouped forms. For me, these forebears offer a form of guidance toward the future.

What I have written in this book represents, at best, an intuitive synthesis of these three frames of reference, even though articulating a clear and actual integration of these theories is, regrettably, outside the scope of this book, and perhaps outside the scope of what life remains to me. But I hope that one day someone else will find it useful to bring together these three theoretical perspectives, to show their areas of overlap and difference, and to demonstrate how each fills in areas missing in the other two. Such an integration would be a highly useful tool for those coming next to the project of studying ourselves in collective life and of figuring out how to work more effectively in groups toward the aspects of life that matter most.

May this volume contribute toward that aspiration.

REFERENCES

Allen, J. R., & Hammond, D. (2003). Groups within groups: Fractals and the successes and failure of a child inpatient psychiatric unit. *Transactional Analysis Journal, 33*, 302–314.

Altorfer, O. (1977). Group dynamics: Dealing with agitation in industry groups. *Transactional Analysis Journal, 7*, 168–169.

Alvarez, A. (2005). *The writer's voice.* New York, NY: Norton.

Apprey, M. (2014). A pluperfect errand: A turbulent return to beginnings in the transgenerational transmission of destructive aggression. *Free Associations, 66*, 16–29.

Apprey, M. (2017). Representing, theorizing and reconfiguring the concept of transgenerational haunting in order to facilitate healing. In S. Grand, & J. Salberg (Eds.), *Trans-generational trauma and the other: Dialogues across history and difference* (pp. 16–37). London: Routledge.

Arendt, H. (1958). *The human condition.* Chicago, IL: University of Chicago Press.

Arendt, H. (1978). Organized guilt and universal responsibility. In R. H. Feldman Ed., *The Jew as pariah: Jewish identity and politics in the modern age* (pp. 225–236). New York, NY: Grove Press. (Original work published 1945).

Arnold, T. J., & Simpson, R. L. (1975). The effects of a TA group on emotionally disturbed school-age boys. *Transactional Analysis Journal, 5*, 238–241.

Balling, R. (2005). Diagnosis of organizational cultures. *Transactional Analysis Journal, 35*, 313–320.

Barrow, G. (2011). Educator as cultivator. *Transactional Analysis Journal, 41*, 308–314.

Barrow, G. (2018). A body of knowledge: Somatic and environmental impacts in the educational encounter. *Transactional Analysis Journal, 48*, 7–17.

Barrow, G., & Newton, T. (Eds.). (2013). *Walking the talk: How transactional analysis is improving behaviour and raising self-esteem.* London: Routledge. (Original work published 2004).

Barrow, G., & Newton, T. (Eds.). (2016). *Educational transactional analysis: An international guide to theory and practice.* London: Routledge.

Berg, J. M., Tymoczko, J. L., & Stryer, L. (2002). *Biochemistry* (5th ed.). New York, NY: Freeman.

Berne, E. (1947). *The mind in action.* New York, NY: Simon and Schuster.

Berne, E. (1949). Some oriental mental hospitals. *American Journal of Psychiatry, 106*, 376–383.

Berne, E. (1950). Cultural aspects of a multiple murder. *Psychiatric Quarterly,* (Supplement), *24*, 250–269.

Berne, E. (1953). Principles of group psychotherapy. *Indian Journal of Neurology & Psychiatry, 4*, 119–137.

REFERENCES

Berne, E. (1954). The natural history of a spontaneous therapy group. *International Journal of Group Psychotherapy, 4*, 74–85.

Berne, E. (1955). Group attendance: Clinical and theoretical considerations. *International Journal of Group Psychotherapy, 5*, 392–403.

Berne, E. (1956). Comparative psychiatry and tropical psychiatry. *American Journal of Psychiatry, 113*, 193–200.

Berne, E. (1958a). Group therapy abroad. *International Journal of Group Psychotherapy, 8*, 466–470.

Berne, E. (1958b). Transactional analysis: A new and effective method of group therapy. *American Journal of Psychotherapy, 12*, 735–743.

Berne, E. (1959a). Difficulties of comparative psychiatry: The Fiji Islands. *American Journal of Psychiatry, 116*, 104–109.

Berne, E. (1959b). The mythology of dark and fair: Psychiatric use of folklore. *Journal of American Folklore, 72*(283), 1–13.

Berne, E. (1959c). Psychiatric epidemiology of the Fiji islands. *Progress in Psychotherapy, 4*, 310–313.

Berne, E. (1960a). The cultural problem: Psychopathology in Tahiti. *American Journal of Psychiatry, 116*, 1076–1081.

Berne, E. (1960b). "Psychoanalytic" versus "dynamic" group therapy. *International Journal of Group Psychotherapy, 10*, 98–103.

Berne, E. (1961a). Cultural factors in group therapy. *International Mental Health Research Newsletter, 3*, 3–4.

Berne, E. (1961b). *Transactional analysis in psychotherapy: A systematic individual and social psychiatry.* New York, NY: Grove Press.

Berne, E. (1962). Teaching group therapy. *Transactional Analysis Bulletin, 1*(2), 11.

Berne, E. (1963). *The structure and dynamics of organizations and groups.* Philadelphia, PA: Lippincott.

Berne, E. (1964). *Games people play: The psychology of human relationships.* New York, NY: Grove Press.

Berne, E. (1966). *Principles of group treatment.* New York, NY: Oxford University Press.

Berne, E. (1968a). *A layman's guide to psychiatry and psychoanalysis* (3rd ed. rev.). New York, NY: Simon and Schuster. (Original work published 1947 as *The mind in action*).

Berne, E. (1968b). Staff-patient staff conferences. *American Journal of Psychiatry, 125*, 286–293.

Berne, E. (1972). *What do you say after you say hello? The psychology of human destiny.* New York, NY: Grove Press.

Berne, E. (1977a). Concerning the nature of diagnosis. In E. Berne (Eds.), *Intuition and ego states: The origins of transactional analysis* (P. McCormick, Ed.) (pp. 33–65). San Francisco, CA: Harper & Row. (Original work published 1952).

Berne, E. (1977b). Eric Berne as group therapist: A verbatim. In M. James, & Contributors (Eds.), *Techniques in transactional analysis for psychotherapists and counselors* (pp. 333–340). Reading, MA: Addison-Wesley. (Original work published 1970).

Berne, E. (1977c). *Intuition and ego states: The origins of transactional analysis* (P. McCormick, Ed.). San Francisco, CA: Harper & Row.

Berne, E. (1977d). The nature of intuition. In E. Berne (Eds.), *Intuition and ego states: The origins of transactional analysis* P. McCormick, Ed.). (pp. 1–31). San Francisco, CA: Harper & Row. (Original work published 1949).

Bernstein, E. L. (1939). Psychiatry in Syria. *American Journal of Psychiatry, 95*, 1415–1419.

REFERENCES

Bion, W. R. (1967a). Attacks on linking. In W. R. Bion (Eds.), *Second thoughts: Selected papers on psycho-analysis* (pp. 93–109). Northvale, NJ: Jason Aronson. (Original work published 1959).
Bion, W. R. (1967b). Commentary. In W. R. Bion (Eds.), *Second thoughts: Selected papers on psycho-analysis* (pp. 120–166). Northvale, NJ: Jason Aronson.
Bion, W. R. (1967c). A theory of thinking. In W. R. Bion (Eds.), *Second thoughts: Selected papers on psycho-analysis* (pp. 110–119). Northvale, NJ: Jason Aronson. (Original work published 1962).
Bion, W. R. (1969). *Experiences in groups and other papers.* New York, NY: Basic Books. (Original work published 1959).
Bion, W. R. (1977a). Attention and interpretation. In W. F. Bion (Eds.), *Seven servants: Four works by Wilfred R. Bion* (pp. 1–136). New York, NY: Jason Aronson. (Original work published 1970).
Bion, W. R. (1977b). Elements of psycho-analysis. In W. F. Bion (Eds.), *Seven servants: Four works by Wilfred R. Bion* (pp. 1–110). New York, NY: Jason Aronson. (Original work published 1963).
Bion, W. R. (1977c). Learning from experience. In W. F. Bion (Eds.), *Seven servants: Four works by Wilfred R. Bion* (pp. 1–111). New York, NY: Jason Aronson. (Original work published 1962).
Bion, W. R. (1980). *Bion in New York and São Paulo* (F. Bion, Ed.). Perthshire: Clunie Press.
Bion, W. R. (2005). *The Tavistock seminars* (F. Bion, Ed.). London: Karnac Books. (Transcripts of lectures given at Tavistock Clinic from 1976-1979).
Blacklidge, V. Y. (1979). What do you do when ... You forget to renew a no suicide contract and the person is improved? *Transactional Analysis Journal, 9,* 193.
Blakeney, R. (1978a). A TA, systems, and organizations approach to team building. *Transactional Analysis Journal, 8,* 158–160.
Blakeney, R. (1978b). A TA-system model for management. *Transactional Analysis Journal, 8,* 259.
Blakeney, R. (1983). The organizational, group and individual levels of analysis in organizational behavior. *Transactional Analysis Journal, 13,* 58–64.
Boholst, F. A. (2003). Effects of transactional analysis group therapy on ego states and ego state perception. *Transactional Analysis Journal, 33,* 254–261.
Bollas, C. (1987). *The shadow of the object: Psychoanalysis of the unthought known.* New York, NY: Columbia University Press.
Bollas, C. (2009). *The evocative object world.* New York, NY: Routledge.
Bonds-White, F. (2003). Anatomy of a training group. *Transactional Analysis Journal, 33,* 344–349.
Bonds-White, F., & Cornell, B. (2002). Perspectives on groups and transactional analysis. *The Script, 32*(5), 1–2.
Bowen, M. (1988). Epilogue: An odyssey toward science. In M. E. Kerr, & M. Bowen (Eds.), *Family evaluation: An approach based on Bowen theory* (pp. 339–386). New York, NY: Norton.
Bowen, M. (1989, January). Unpublished memorandum. Washington, DC: The Georgetown Family Center.
Bowen, M. (1991). Diversity from unity. In J. E. Earley (Ed.), *Individuality and cooperative action* (pp. 87–90). Washington, DC: Georgetown University Press.
Bowen, M. (1994). *Family therapy in clinical practice.* Northvale, NJ: Jason Aronson. (Original work published 1978).
Bowen, M. (1995). A psychological formulation of schizophrenia [drafted 1955–1957; intro. by C. Rakow]. *Family Systems, 2,* 17–47.
Brooks, D. (2009, 27 November). The other education. *The New York Times,* p. A39.

REFERENCES

Brown, M. (1974). TA and community consultation. *Transactional Analysis Journal, 4*(1), 20–22.

Bushe, G. R., & Marshak, R. J. (2009). Revisioning organization development: Diagnostic and dialogic premises and patterns of practice. *Journal of Applied Behavioral Science, 45,* 348–368.

Bushe, G. R., & Marshak, R. J. (2014). The dialogic mindset in organization development. *Research in Organization Change and Development, 22,* 55–97.

Caizzi, C., & Giacometto, R. (2008). Working with unconscious processes in a short-term transactional analysis group. *Transactional Analysis Journal, 38,* 164–170.

Campos, L. P. (1971). Transactional analysis group leadership operations. *Transactional Analysis Journal, 1*(4), 21–24.

Campos, L. P. (2012). Cultivating cultures of courage with transactional analysis. *Transactional Analysis Journal, 42,* 209–219.

Caper, R. (1999). *A mind of one's own: A Kleinian view of self and object.* London: Routledge.

Capoferri, C. (2014). The collective parent: Theory and process. *Transactional Analysis Journal, 44,* 175–185.

Cardon, A. (1993). Organizational holistics. *Transactional Analysis Journal, 23,* 66–69.

Cassoni, E., & Filippi, S. (2013). Travel mates: Transactional analysis groups for couples. *Transactional Analysis Journal, 43,* 334–346.

Chandran, S. (2007). Connecting with the guru within: Supervision in the Indian context. *Transactional Analysis Journal, 37,* 218–226.

Chinnock, K. (2011). Relational transactional analysis supervision. *Transactional Analysis Journal, 41,* 336–350.

Christensen, B. (2005). Emerging participative exploration: Consultation as research. In R. Stacey, & D. Griffin (Eds.), *A complexity perspective on researching organizations: Taking experience seriously* (pp. 78–106). London: Routledge.

Clarke, J. I. (1981). Differences between special fields and clinical groups. *Transactional Analysis Journal, 11,* 169–170.

Clarkson, P. (1991). Group imago and the stages of group development. *Transactional Analysis Journal, 21,* 36–50.

Clarkson, P. (1992). *Transactional analysis psychotherapy: An integrated approach.* London: Routledge.

Cornell, W. F. (2008). A community for thinking. In W. F. Cornell (Eds.), *Explorations in transactional analysis: The Meech Lake papers* (pp. 265–267). Pleasanton, CA: TA Press.

Cornell, W. F. (2010). Aspiration or adaptation? An unresolved tension in Eric Berne's basic beliefs. *Transactional Analysis Journal, 40,* 243–253.

Cornell, W. F. (2013). "Special fields": A brief history of an anxious dilemma and its lingering consequences for transactional analysis counselors. *Transactional Analysis Journal, 43,* 7–13.

Cornell, W. F. (2016). In conflict and community: A century of turbulence working and living in groups. *Transactional Analysis Journal, 46,* 136–148.

Cornell, W. F. (2018). If it is not for all, it is not for us: Reflections on racism, nationalism, and populism in the United States. *Transactional Analysis Journal, 48,* 97–110.

Cornell, W. F. (2019). Fostering freedom for play, imagination, and uncertainty in professional learning environments. In W. F. Cornell (Eds.), *At the interface of transactional analysis, psychoanalysis, and body psychotherapy* (pp. 33–38). London: Routledge.

REFERENCES

Cornell, W. F., & Bonds-White, F. (Eds.). (2003). Transactional analysis and groups [Theme issue]. *Transactional Analysis Journal, 33*(4).

Cornell, W. F., & Hargaden, H. (Eds.). (2005). *From transactions to relations: The emergence of a relational tradition in transactional analysis.* London: Haddon Press.

Cornell, W. F., & Hargaden, H. (Eds.). (2020). *The evolution of a relational paradigm in transactional analysis: What's the relationship got to do with it?* London: Routledge.

Cornell, W. F., & Landaiche, N. M., III. (2006). Impasse and intimacy: Applying Berne's concept of script protocol. *Transactional Analysis Journal, 36*, 196–213.

Cory, T. L., & Page, D. (1978). Playing games in group therapy. *Transactional Analysis Journal, 8*, 229.

Cox, M. (2007). On doing supervision. *Transactional Analysis Journal, 37*, 104–114.

Crespelle, I. (1988). An active observation model for therapists in training. *Transactional Analysis Journal, 18*, 249–253.

Dalal, F. (2016). The individual and the group: The twin tyrannies of internalism and individualism. *Transactional Analysis Journal, 46*, 88–100.

Damasio, A. (2010). *Self comes to mind: Constructing the conscious brain.* New York, NY: Pantheon Books.

Darwin, C. (1872). *The expression of the emotions in man and animals.* London: John Murray.

de Graaf, A. (2013). The group in the individual: Lessons learned from working with and in organizations and groups. *Transactional Analysis Journal, 43*, 311–320.

de Graaf, A., & Levy, J. (2016). Are transactional analysis training program groups sufficiently disturbing? *Transactional Analysis Journal, 46*, 222–231.

Deaconu, D. (2013). The group quest: Searching for the group inside me, inside you, and inside the community. *Transactional Analysis Journal, 43*, 291–295.

Dewey, J. (1981). *The later works, 1925-1953* (Vol. 1: 1925: Experience and nature, J. A. Boydston (Ed.)). Carbondale: Southern Illinois University Press, (Original work published 1925).

Diamond, J. (1987, May). The worst mistake in the history of the human race. *Discover Magazine*, 64–66.

Dickinson, E. (1976). It is easy to work when the soul is at play. In E. Dickinson (Eds.), *The complete poems of Emily Dickinson* (T. H. Johnson, Ed.) (pp. 111–112). Boston, MA: Back Bay Books. (Poem written circa 1861, originally published 1945.

Douglas, M. (1986). *How institutions think.* Syracuse, NY: Syracuse University Press.

Downing, G. (2008). A different way to help. In A. Fogel, B. J. King, & S. G. Shanker (Eds.), *Human development in the twenty-first century: Visionary ideas from systems scientists* (pp. 200–205). Cambridge: Cambridge University Press.

Drego, P. (1996). Cultural parent oppression and regeneration. *Transactional Analysis Journal, 26*, 58–77.

Drye, B. (1980). Psychoanalytic definitions of cure: Beyond contract completion. *Transactional Analysis Journal, 10*, 124–130.

Eigen, M. (1998). *The psychoanalytic mystic.* Binghamton, NY: Esf Publishers.

Eisenberger, N. I., Jarcho, J. M., Lieberman, M. D., & Naliboff, B. D. (2006). An experimental study of shared sensitivity to physical pain and social rejection. *Pain, 126*, 132–138.

Eisenberger, N. I., & Lieberman, M. D. (2005). Why it hurts to be left out: The neurocognitive overlap between physical and social pain. In K. D. Williams, J. P. Forgas, & W. von Hippel (Eds.), *The social outcast: Ostracism, social exclusion, rejection, and bullying* (pp. 109–127). New York, NY: Psychology Press.

Eisenberger, N. I., Lieberman, M. D., & Williams, K. D. (2003). Does rejection hurt? An fMRI study of social exclusion. *Science, 302*, 290–292.

Eliot, G. (1961). *The mill on the floss.* New York, NY: Harcourt, Brace and World. (Original work published 1860).

Emmerton, N., & Newton, T. (2004). The journey of educational transactional analysis from its beginnings to the present. *Transactional Analysis Journal, 34,* 283–291.

English, F. (1975a). Shame and social control. *Transactional Analysis Journal, 5,* 24–28.

English, F. (1975b). The three-cornered contract. *Transactional Analysis Journal, 5,* 383–384.

English, F. (1994). Shame and social control revisited. *Transactional Analysis Journal, 24,* 109–120.

English, F. (1977). What shall I do tomorrow? Reconceptualizing transactional analysis. In G. Barnes (Ed.), *Transactional analysis after Eric Berne: Teachings and practices of three TA schools* (pp. 287–347). New York, NY: Harper's College Press.

Ernst, F. H., Jr. (1971a). The OK corral: The grid for get-on-with. *Transactional Analysis Journal, 1*(4), 33–42.

Ernst, F. H., Jr. (Ed.). (1971b). Featuring—Education [Theme issue]. *Transactional Analysis Journal, 1*(4).

Erskine, R. G. (1994). Shame and self-righteousness: Transactional analysis perspectives and clinical interventions. *Transactional Analysis Journal, 24,* 86–102.

Erskine, R. G. (2009). The culture of transactional analysis: Theory, methods, and evolving patterns. *Transactional Analysis Journal, 39,* 14–21.

Erskine, R. G. (2013). Relational group process: Developments in a transactional analysis model of group psychotherapy. *Transactional Analysis Journal, 43,* 262–275.

Erskine, R. G., Clark, B., Evans, K. R., Goldberg, C., Hyams, H., James, S., & O'Reilly-Knapp, M. (1994). The dynamics of shame: A roundtable discussion. *Transactional Analysis Journal, 24,* 80–85.

Fanon, F. (1967). *Black skin, white masks* (C. L. Markmann, Trans.). New York, NY: Grove Press. (Original work published 1952).

Fast, I. (2006). A body-centered mind: Freud's more radical idea. *Contemporary Psychoanalysis, 42*(2), 273–295.

Fowlie, H., & Sills, C. (Eds.). (2011). *Relational transactional analysis: Principles in practice.* London: Karnac Books.

Fox, L. A., & Baker, K. G. (2009). *Leading a business in anxious times: A systems-based approach to becoming more effective in the workplace.* Chicago, IL: Care Communications Press.

Freire, P. (2000). *Pedagogy of the oppressed: 30th anniversary edition* (M. Bergman Ramos, Trans.). New York, NY: Continuum. (Original work published 1970; translation of original 1968 Portuguese manuscript.

Furness, J. (2006). *The enteric nervous system.* Malden, MA: Blackwell Publishing.

Gellert, S. D., & Wilson, G. (1978). Contracts. *Transactional Analysis Journal, 8,* 10–15.

Gershon, M. D. (1998). *The second brain: The scientific basis of gut instinct.* New York, NY: HarperCollins.

Gibson, D. L. (1974). The classic non-contract. *Transactional Analysis Journal, 4*(2), 31.

Gilligan, J. (2000). *Violence: Reflections on our deadliest epidemic.* London: Jessica Kingsley Publishers. Original work published 1996.

Grant, J. (2004). How the philosophical assumptions of transactional analysis complement the theory of adult education. *Transactional Analysis Journal, 34,* 272–276.

REFERENCES

Gray, L. (2015, June 10). Method reveals what bacteria sense in their surroundings: Knowing how environmental signals modulate bacterial behavior could help combat biofouling and antibiotic resistance. *HSNewsBeat* [University of Washington Health Sciences]. Retrieved November 20, 2019, from https://news room.uw.edu/story/method-reveals-what-bacteria-sense-their-surroundings

Grégoire, J. (2015). Thinking, theory, and experience in the helping professions: A phenomenological description. *Transactional Analysis Journal, 45*, 59–71.

Groder, M. (Ed.). (1975). Organizations [Special issue]. *Transactional Analysis Journal, 5*(4).

Gurowitz, E. M. (1975). Group boundaries and leadership potency. *Transactional Analysis Journal, 5*, 183–185.

Hacking, I. (2012). Introductory essay. In T. S. Kuhn (Eds.), *The structure of scientific revolutions* (4th ed., pp. vii–xxxvii). Chicago, IL: University of Chicago Press.

Haimowitz, M. L. (1973). Short term contracts. *Transactional Analysis Journal, 3*(2), 34.

Haimowitz, M. L. (1975). Training and therapy in large groups without charge. *Transactional Analysis Journal, 5*, 36–37.

Harari, Y. N. (2015). *Sapiens: A brief history of humankind* (Y. Harari, J. Purcell, & H. Watzman, Trans.). New York, NY: Harper. (Original work published 2011).

Hargaden, H. (2013). Building resilience: The role of firm boundaries and the third in relational group therapy. *Transactional Analysis Journal, 43*, 284–290.

Hargaden, H., & Fenton, B. (2005). An analysis of nonverbal transactions drawing on theories of subjectivity. *Transactional Analysis Journal, 35*, 173–186.

Hargaden, H., & Sills, C. (2002a). Group psychotherapy. In H. Hargaden, & C. Sills (Eds.), *Transactional analysis: A relational perspective* (pp. 139–152). Hove, England: Brunner-Routledge.

Hargaden, H., & Sills, C. (2002b). *Transactional analysis: A relational perspective*. Hove, England: Brunner-Routledge.

Harley, K. (2006). A lost connection: Existential positions and Melanie Klein's infant development. *Transactional Analysis Journal, 36*, 252–269.

Hawkes, L. (2003). The tango of therapy: A dancing group. *Transactional Analysis Journal, 33*, 288–301.

Hawkes, L. (2019). Tango and TA in Bulgaria. *The Script, 49*(5), 11–12.

Hay, J. (2000). Organizational transactional analysis: Some opinions and ideas. *Transactional Analysis Journal, 30*, 223–232.

Hay, J. (2012). *Transactional analysis for trainers* (2nd ed.). Hertford, England: Sherwood Publishing.

Hinshelwood, R. D. (1983). Editorial: Our three-way see-saw. *International Journal of Therapeutic Communities, 4*, 167–168.

Hinshelwood, R. D. (1987). *What happens in groups: Psychoanalysis, the individual and the community*. London: Free Association Books.

Hopkin, M. (2008, May 8). Bacteria "can learn": Colonies evolve to anticipate changes in their surroundings. *Nature*. Retrieved November 20, 2019, from www.nature.com/news/2008/080508/full/news.2007.360.html

Hyams, H. (1994). Shame: The enemy within. *Transactional Analysis Journal, 24*, 255–264.

Jackson, P. L., Meltzoff, A. N., & Decety, J. (2005). How do we perceive the pain of others? A window into the neural processes involved in empathy. *Neuroimage, 24*, 771–779.

Jacobs, A. (1991). Autocracy: Groups, organizations, nations, and players. *Transactional Analysis Journal, 21*, 199–206.

James, J. (Ed.). (1983). Cultural scripts [Theme issue]. *Transactional Analysis Journal, 13*(4).

REFERENCES

James, M. (Ed.). (1979). [Issue featuring education]. *Transactional Analysis Journal, 9*(4).

James, N. L. (1993). Addiction in society: Shame, addiction and the child ego. In N. James (Ed.), *The Minneapolis papers: Selections from the 31st annual ITAA conference* (pp. viii–xi). Madison, WI: Omnipress.

James, N. L. (1994). Cultural frame of reference and intergroup encounters: A TA approach. *Transactional Analysis Journal, 24,* 206–210.

Janis, I. L. (1971, November). Groupthink. *Psychology Today, 5,* 43–46, 74–76.

Johnson, M. (2017). *Embodied mind, meaning, and reason: How our bodies give rise to understanding.* Chicago, IL: University of Chicago Press.

Johnson, M., & Rohrer, T. (2017). We are live creatures: Embodiment, American pragmatism, and the cognitive organism. In M. Johnson (Eds.), *Embodied mind, meaning, and reason: How our bodies give rise to understanding* (pp. 67–97). Chicago, IL: University of Chicago Press.

Joseph, M. R. (2012). Therapeutic operations can be educational operations too. *Transactional Analysis Journal, 42,* 110–117.

Kapur, R., & Miller, K. (1987). A comparison between therapeutic factors in TA and psychodynamic therapy groups. *Transactional Analysis Journal, 17,* 294–300.

Kinoy, B. P. (1985). Self-help groups in the management of anorexia nervosa and bulimia: A theoretical base. *Transactional Analysis Journal, 15,* 73–78.

Kolb, D. A. (1984). *Experiential learning: Experience as the source of learning and development.* Englewood Cliffs, NJ: Prentice Hall.

Krausz, R. R. (1986). Power and leadership in organizations. *Transactional Analysis Journal, 16,* 85–94.

Krausz, R. R. (1996). Transactional analysis and the transformation of organizations. *Transactional Analysis Journal, 26,* 52–57.

Krausz, R. R. (2013). Living in groups. *Transactional Analysis Journal, 43,* 366–374.

Kreyenberg, J. (2005). Transactional analysis in organizations as a systemic constructivist approach. *Transactional Analysis Journal, 35,* 300–310.

Kuechler, C. F., & Andrews, J. (1996). Providing a bridge between psychoeducational groups and social work. *Transactional Analysis Journal, 26,* 175–181.

Kuhn, T. S. (2012a). Postscript—1969. In T. S. Kuhn (Eds.), *The structure of scientific revolutions* (4th ed., pp. 173–208). Chicago, IL: University of Chicago Press. (Original work published 1969).

Kuhn, T. S. (2012b). *The structure of scientific revolutions* (4th ed.). Chicago, IL: University of Chicago Press. (Original work published 1962).

Lakoff, G., & Johnson, M. (1999). *Philosophy in the flesh: The embodied mind and its challenge to western thought.* New York, NY: Basic Books.

Landaiche, N. M., III. (2005). Engaged research: Encountering a transactional analysis training group through Bion's concept of containing. *Transactional Analysis Journal, 35,* 147–156.

Landaiche, N. M., III. (2007). Skepticism and compassion in human relations work. *Transactional Analysis Journal, 37,* 17–31.

Landaiche, N. M., III. (2009). Understanding social pain dynamics in human relations. *Transactional Analysis Journal, 39,* 229–238.

Landaiche, N. M., III. (2010). The importance of interdisciplinary training experiences. *The Script, 40*(5), 1, 7.

Landaiche, N. M., III. (2012). Learning and hating in groups. *Transactional Analysis Journal, 42,* 186–198.

Landaiche, N. M., III. (2013). Looking for trouble in groups developing the professional's capacity. *Transactional Analysis Journal, 43,* 296–310.

Landaiche, N. M., III. (2014). Failure and shame in professional practice: The role of social pain, the haunting of loss. *Transactional Analysis Journal, 44,* 268–278.

Landaiche, N. M., III. (2015, 5 August). *Preparing the next generation: A 7-year study of a seminar on Bowen theory.* Presentation given at the Western Pennsylvania Family Center 1st International Conference on Bowen Family Systems Theory, Pittsburgh, PA.

Landaiche, N. M., III. (2016a). Maturing as a community effort: A discussion of Dalal's and Samuels's perspectives on groups and individuals. *Transactional Analysis Journal, 46,* 116–120.

Landaiche, N. M., III. (2016b). [Review of the book *Educational transactional analysis: An international guide to theory and practice* by Giles Barrow and Trudi Newton, Eds.]. *Transactional Analysis Journal, 46,* 249–251.

Lawrence, W. G., & Armstrong, D. (1998). Destructiveness and creativity in organizational life: Experiencing the psychotic edge. In P. B. Talamo, F. Borgogno, & S. A. Merciai (Eds.), *Bion's legacy to groups* (pp. 53–68). London: Karnac Books.

Lee, A. (1997). Process contracts. In C. Sills (Ed.), *Contracts in counselling* (pp. 94–112). London: Sage.

Lee, A. (2014). The development of a process group. *Transactional Analysis Journal, 44,* 41–52.

Lerkkanen, M.-K., & Temple, S. (2004). Student teachers' professional and personal development through academic study of educational transactional analysis. *Transactional Analysis Journal, 34,* 253–271.

Lewin, K. (1948). *Resolving social conflicts: Selected papers on group dynamics.* (G. W. Lewin, Ed.). New York, NY: Harper & Row.

Lewin, M. (2000). "I'm not talking to you": Shunning as a form of violence. *Transactional Analysis Journal, 30,* 125–131.

Lewis, T., Amini, F., & Lannon, R. (2001). *A general theory of love.* New York, NY: Vintage Books-Random House.

Lieberman, M. D., & Eisenberger, N. I. (2005). A pain by any other name (rejection, exclusion, ostracism) still hurts the same: The role of dorsal anterior cingulate cortex in social and physical pain. In J. T. Cacioppo, P. Visser, & C. Pickett (Eds.), *Social neuroscience: People thinking about people* (pp. 167–187). Cambridge, MA: MIT Press.

Ligabue, S. (2007). Being in relationship: Different languages to understand ego states, script, and the body. *Transactional Analysis Journal, 37,* 294–306.

Loomis, M. (1982). Contracting for change. *Transactional Analysis Journal, 12,* 51–55.

MacDonald, G., Kingsbury, R., & Shaw, S. (2005). Adding insult to injury: Social pain theory and response to social exclusion. In K. D. Williams, J. P. Forgas, & W. von Hippel (Eds.), *The social outcast: Ostracism, social exclusion, rejection, and bullying* (pp. 77–90). New York, NY: Psychology Press.

MacLean, P. D. (1990). *The triune brain in evolution: Role in paleocerebral functions.* New York, NY: Plenum Press.

MacLean, P. D. (1993). Cerebral evolution of emotion. In M. Lewis, & J. M. Haviland (Eds.), *Handbook of emotions* (pp. 67–83). New York, NY: Guilford.

Manor, O. (1992). Transactional analysis, object relations, and the systems approach: Finding the counterparts. *Transactional Analysis Journal, 22,* 4–15.

Maquet, J. (2012). From psychological contract to frame dynamics: Between light and shadow. *Transactional Analysis Journal, 42,* 17–27.

Marcum, J. A. (2015). *Thomas Kuhn's revolutions: A historical and an evolutionary philosophy of science?* London: Bloomsbury.

Mazzetti, M. (2010). Eric Berne and cultural script. *Transactional Analysis Journal, 40,* 187–195.

REFERENCES

Mazzetti, M. (2012). Phantoms in organizations. *Transactional Analysis Journal, 42,* 199–208.

McGrath, G. (1994). Ethics, boundaries, and contracts: Applying moral principles. *Transactional Analysis Journal, 24,* 6–14.

McQuillin, J., & Welford, E. (2013). How many people are gathered here? Group work and family constellation theory. *Transactional Analysis Journal, 43,* 352–365.

Mellor, K. (2007). [Letter to the editor]. *Transactional Analysis Journal, 37,* 174–176.

Micholt, N. (1992). Psychological distance and group interventions. *Transactional Analysis Journal, 22,* 228–233.

Milnes, P. (2019). "Written on my heart in burning letters": Putting soul and spirit into a transcendent physis. *Transactional Analysis Journal, 49,* 144–157.

Milosz, C. (1975). My intention. In C. Milosz (Eds.), *Visions from San Francisco Bay* (R. Lourie, Trans.) (pp. 3–5). New York, NY: Farrar, Straus and Giroux. (Original work published 1969.

Minnich, E. K. (2003). Teaching thinking: Moral and political considerations. *Change, 35*(5), 18–24.

Misel, L. T. (1975). Stages of group treatment. *Transactional Analysis Journal, 5,* 385–391.

Moiso, C. (1976). The contract card. *Transactional Analysis Journal, 6,* 298–299.

Morrison, T. (1988). *Beloved.* New York, NY: Plume.

Mothersole, G. (1996). Existential realities and no-suicide contracts. *Transactional Analysis Journal, 26,* 151–159.

Mountain, A., & Davidson, C. (2005). Assessing systems and processes in organizations. *Transactional Analysis Journal, 35,* 336–345.

Napper, R., & Newton, T. (2014). *Tactics: Transactional analysis concepts for all trainers, teachers, and tutors + insight into collaborative learning strategies.* Ipswich, England: TA Resources. Original work published 2000.

Newton, T. (2003). Identifying educational philosophy and practice through imagoes in transactional analysis training groups. *Transactional Analysis Journal, 33,* 321–331.

Newton, T. (Ed.). (2004). Transactional analysis and education [Theme issue]. *Transactional Analysis Journal, 34*(3).

Newton, T. (2006). Script, psychological life plans, and the learning cycle. *Transactional Analysis Journal, 36,* 186–195.

Newton, T. (2011a). The nature and necessity of risk: Minding the gap in education. *Transactional Analysis Journal, 41,* 114–117.

Newton, T. (2011b). Transactional analysis now: Gift or commodity? *Transactional Analysis Journal, 41,* 315–321.

Newton, T. (2012). The supervision triangle: An integrating model. *Transactional Analysis Journal, 42,* 103–109.

Newton, T. (2014). Learning imagoes update. *Transactional Analysis Journal, 44,* 31–40.

Newton, T., & Napper, R. (Eds.). (2009). Transactional analysis training [Theme issue]. *Transactional Analysis Journal, 39*(4).

Noce, J. (1978). A model for the collective parenting function of therapeutic communities. *Transactional Analysis Journal, 8,* 332–338.

Noriega, G. (2010). The transgenerational script of transactional analysis. *Transactional Analysis Journal, 40,* 196–204.

Novellino, M. (2005). Transactional psychoanalysis: Epistemological foundations. *Transactional Analysis Journal, 35,* 157–172.

Novellino, M. (2010). The demon and sloppiness: From Berne to transactional psychoanalysis. *Transactional Analysis Journal, 40,* 288–294.

REFERENCES

Nuttall, J. (2000). Intrapersonal and interpersonal relations in management organizations. *Transactional Analysis Journal, 30*, 73–82.

Nykodym, N. (1978). Transactional analysis: A strategy for the improvement of supervisory behavior. *Transactional Analysis Journal, 8*, 254–258.

Nykodym, N., Freedman, L. D., Simonetti, J. L., Nielsen, W. R., & Battles, K. (1995). Mentoring: Using transactional analysis to help organizational members use their energy in more productive ways. *Transactional Analysis Journal, 25*, 170–179.

O'Hearne, J. J. (1977). The patient as collaborator. In M. James, & Contributors (Eds.), *Techniques in transactional analysis for psychotherapists and counselors* (pp. 176–184). Reading, MA: Addison-Wesley.

O'Reilly-Knapp, M. (Ed.). (1994). Shame [Theme issue]. *Transactional Analysis Journal, 24*(2).

Oates, S. (2010). The indomitable spirit of Berne and Cohen: "If you can't do it one way, try another". *Transactional Analysis Journal, 40*, 300–304.

Ogden, T. H. (2005). *This art of psychoanalysis: Dreaming undreamt dreams and interrupted cries*. London: Routledge.

Panksepp, J. (1998). *Affective neuroscience: The foundations of human and animal emotions*. New York, NY: Oxford University Press.

Peck, H. B. (1978). Integrating transactional analysis and group process approaches in treatment. *Transactional Analysis Journal, 8*, 328–331.

Petriglieri, G. (2004). What do you want to learn tomorrow? *The Script, 34*(4), 1, 7.

Petriglieri, G. (2010, 15 August). *Respected marginality* [Closing keynote speech]. International Transactional Analysis Association Conference, Montreal, Canada.

Petriglieri, G., & Wood, J. D. (2003). The invisible revealed: Collusion as an entry to the group unconscious. *Transactional Analysis Journal, 33*, 332–343.

Piccinino, G. (2018). Reflections on physis, happiness, and human motivation. *Transactional Analysis Journal, 48*, 272–285.

Poindexter, W. R. (1975). Organizational games. *Transactional Analysis Journal, 5*, 379–382.

Pratt, K., & Mbaligontsi, M. (2014). Transactional analysis transforms community care workers in South Africa. *Transactional Analysis Journal, 44*, 53–67.

Ramanujan, K. (2006, May 17). How does the lowly bacterium sense its environment? Cornell researchers discover lattice of supersensitive receptors. *Cornell Chronicle*. Retrieved November 20, 2019, from https://news.cornell.edu/stories/2006/05/researchers-discover-how-bacteria-sense-their-environments

Ranci, D. (2002). Il gruppo di apprendimento. Riflessioni su un'esperienza [A learning group: Thinking over an experience]. *Quaderni di Psicologia, Analisi Transazionale e Scienze Umane*, 35–36, n.p.

Roberts, D. L. (1975). Treatment of cultural scripts. *Transactional Analysis Journal, 5*, 29–35.

Roberts, D. L. (1984). Contracting for peace—The first step in disarmament. *Transactional Analysis Journal, 14*, 229–230.

Robinson, J. (2003). Groups and group dynamics in a therapeutic community. *Transactional Analysis Journal, 33*, 315–320.

Rogers, C. (1967). The process of the basic encounter group. In J. F. T. Bugental (Ed.), *Challenges of humanistic psychology* (pp. 261–276). New York, NY: McGraw-Hill.

Rogers, C. R. (1955). Persons or science? A philosophical question. *American Psychologist, 10*, 267–278.

Samuels, A. (2016). "I rebel, therefore we are" (Albert Camus): New political thinking on individual responsibility for group, society, culture, and planet. *Transactional Analysis Journal, 46*, 101–108.

REFERENCES

Sardello, R. (1971). A reciprocal participation model of experimentation. In A. Giorgi, W. F. Fischer, & R. Von Eckartsberg (Eds.), *Duquesne studies in phenomenological psychology* (Vol. 1, pp. 58–65). Pittsburgh, PA: Duquesne University Press.

Saru, P. K., Cariapa, A., & Manacha, S., with Napper, R. (2009). India as a unique context for developing transactional analysts. *Transactional Analysis Journal, 39*, 326–332.

Schanuel, M. (1976). Don't ingroup yourself out, TA. *Transactional Analysis Journal, 6*, 316–317.

Schein, E. H. (2010). *Organizational culture and leadership* (4th ed.). San Francisco, CA: Jossey-Bass. Original work published 1989.

Schlesinger, W. M. (Ed.). (1978). [Issue featuring education]. *Transactional Analysis Journal, 8*(3).

Schmid, P. F., & O'Hara, M. (2007). Group therapy and encounter groups. In M. Cooper, M. O'Hara, P. F. Schmid, & G. Wyatt (Eds.), *The handbook of person-centred psychotherapy and counselling* (pp. 93–106). Basingstoke, England: Palgrave Macmillan.

Schore, A. N. (2005). Attachment, affect regulation, and the developing right brain: Linking developmental neuroscience to pediatrics. *Pediatrics in Review, 26*, 204–217.

Schur, T. J. (2002). Supervision as a disciplined focus on self and not the other: A different systems model. *Contemporary Family Therapy, 24*, 399–422.

Schur, T. J. (2011). A supervision model based in Bowen theory and language. In O. Cohn Bregman, & C. M. White (Eds.), *Bringing systems thinking to life: Expanding the horizons for Bowen family systems theory* (pp. 281–292). New York, NY: Brunner-Routledge.

Senge, P. M. (1990). *The fifth discipline: The art and practice of the learning organization.* New York, NY: Doubleday.

Senge, P. M., Cambron-McCabe, N., Lucas, T., Smith, B., Dutton, J., & Kleiner, A. (2012). *Schools that learn: A fifth discipline fieldbook for educators, parents, and everyone who cares about education* (Rev ed.). New York, NY: Crown Business. Original work published 2000.

Shaskan, D. A., & Moran, W. L. (1986). Influence of group psychotherapy: A thirty-eight-year follow-up. *Transactional Analysis Journal, 16*, 137–138.

Shaw, P. (2002). *Changing conversations in organizations: A complexity approach to change.* London: Routledge.

Sills, C. (2003). Role lock: When the whole group plays a game. *Transactional Analysis Journal, 33*, 282–287.

Sinclair-Brown, W. (1982). A TA redecision group psychotherapy treatment program for mothers who physically abuse and/or seriously neglect their children. *Transactional Analysis Journal, 12*, 39–45.

Singer, T., Seymour, B., O'Doherty, J., Kaube, H., Dolan, R. J., & Frith, C. D. (2004). Empathy for pain involves the affective but not sensory components of pain. *Science, 303*, 1157–1162.

Solomon, C. (2010). Eric Berne the therapist: One patient's perspective. *Transactional Analysis Journal, 40*, 183–186.

Spence, M. T. (1974). Group work with old people. *Transactional Analysis Journal, 4*(2), 35–37.

Stacey, R., & Griffin, D. (2005). Introduction: Researching organizations from a complexity perspective. In R. Stacey, & D. Griffin (Eds.), *A complexity perspective on researching organizations: Taking experience seriously* (pp. 1–12). London: Routledge.

REFERENCES

Stacey, R. D. (2001). *Complex responsive processes in organizations: Learning and knowledge creation*. London: Routledge.

Steele, C. A., & Porter-Steele, N. (2003). OKness-based groups. *Transactional Analysis Journal, 33*, 276–281.

Steiner, C. M., & Cassidy, W. (1969). Therapeutic contracts in group treatment. *Transactional Analysis Bulletin, 8*(30), 29–31.

Stern, D. H. (1985). *The interpersonal world of the infant: A view from psychoanalysis and developmental psychology*. New York, NY: Basic Books.

Stummer, G. (2002). An update on the use of contracting. *Transactional Analysis Journal, 32*, 121–123.

Stuthridge, J. (Ed.). (2013). Perspectives on working with groups [Theme issue]. *Transactional Analysis Journal, 43*(4).

Stuthridge, J., & Rowland, H. (Eds.). (2019). Transgenerational trauma [Theme issue]. *Transactional Analysis Journal, 49*(4).

Sullivan, H. S. (1953). *The interpersonal theory of psychiatry*. H. S. Perry & M. L. Gawel, Eds. New York, NY: Norton.

Tangolo, A. E. (2015). Group imago and dreamwork in group therapy. *Transactional Analysis Journal, 45*, 179–190.

Tangolo, A. E., & Massi, A. (2018). A contemporary perspective on transactional analysis group therapy. *Transactional Analysis Journal, 48*, 209–223.

Terlato, V. (2017). Secret gardens and dusty roads: Psychological levels and defensiveness in contracting between patient and therapist. *Transactional Analysis Journal, 47*, 8–18.

Thomas, H. E. (1997). *The shame response to rejection*. Sewickley, PA: Albanel Publishers.

Thompson, E. (2011, 24 May). *The vital role of courage in leadership*. Presentation given for the Application of Bowen Theory Series, Western Pennsylvania Family Center, Pittsburgh, PA.

Thomson, S. H. (1974). Insight for the sightless: A TA group for the blind. *Transactional Analysis Journal, 4*(1), 13–17.

Tudor, K. (1991). Children's groups: Integrating TA and gestalt perspectives. *Transactional Analysis Journal, 21*, 12–20.

Tudor, K. (1999). *Group counselling*. London: Sage.

Tudor, K. (2002). Transactional analysis supervision or supervision analyzed transactionally? *Transactional Analysis Journal, 32*, 39–55.

Tudor, K. (2007). Training in the person-centred approach. In M. Cooper, M. O'Hara, P. F. Schmid, & G. Wyatt (Eds.), *The handbook of person-centred psychotherapy and counselling* (pp. 379–389). Basingstoke, England: Palgrave Macmillan.

Tudor, K. (2009). "In the manner of": Transactional analysis teaching of transactional analysts. *Transactional Analysis Journal, 39*, 276–292.

Tudor, K. (2013). Group imago and group development: Two theoretical additions and some ensuing adjustments. *Transactional Analysis Journal, 43*, 321–333.

van Beekum, S. (2006). The relational consultant. *Transactional Analysis Journal, 36*, 318–329.

van Beekum, S. (2012). Connecting with the undertow: The methodology of the relational consultant. *Transactional Analysis Journal, 42*, 126–133.

van Beekum, S. (2013). Changing the focus: The impact of sibling issues on group dynamics. *Transactional Analysis Journal, 43*, 347–351.

van Beekum, S. (2015). A relational approach in consulting: A new formulation of transactional analysis theory in the field of organizations. *Transactional Analysis Journal, 45*, 167–178.

REFERENCES

van Beekum, S., & Laverty, K. (2007). Social dreaming in a transactional analysis context. *Transactional Analysis Journal, 37,* 227–234.

van Poelje, S. (2004). Learning for leadership. *Transactional Analysis Journal, 34,* 223–228.

van Poelje, S. (Ed.). (2005). The evolving field of organizational transactional analysis theme issue. *Transactional Analysis Journal, 35*(4).

Vanwynsberghe, J. (1998). Therapy with alcoholic clients: Guidelines for good contracts. *Transactional Analysis Journal, 28,* 127–131.

Webster's new collegiate dictionary. (1975). Springfield, MA: G. & C. Merriam Co.

Weiss, J. B., & Weiss, L. (1998). Perspectives on the current state of contractual regressive therapy. *Transactional Analysis Journal, 28,* 45–47.

Wells, M. (2002). My journey through groups. *The Script, 32*(5), 6.

Wheatley, M. J. (2017). *Who do we choose to be? Facing reality, claiming leadership, restoring sanity.* Oakland, CA: Berrett-Koehler Publishers.

White, J. D., & White, T. (1975). Cultural scripting. *Transactional Analysis Journal, 5,* 12–23.

White, T. (1999). No-psychosis contracts. *Transactional Analysis Journal, 29,* 133–138.

White, T. (2001). The contact contract. *Transactional Analysis Journal, 31,* 194–198.

Whyte, W. H., Jr. (1952, March). Groupthink. *Fortune,* 114–117, 142, 146.

Williams, K. D., Forgas, J. P., von Hippel, W., & Zadro, L. (2005). The social outcast: An overview. In K. D. Williams, J. P. Forgas, & W. von Hippel (Eds.), *The social outcast: Ostracism, social exclusion, rejection, and bullying* (pp. 1–16). New York, NY: Psychology Press.

Winnicott, D. W. (1975). Anxiety associated with insecurity. In D. W. Winnicott (Eds.), *Through paediatrics to psycho-analysis* (pp. 97–100). London: Karnac Books. (Paper presented 1952, original work published 1958).

Woods, K. (2007). Surrender as a group norm. *Transactional Analysis Journal, 37,* 235–239.

Wright, A. L. (1977). The three-cornered contract revisited. *Transactional Analysis Journal, 7,* 216.

INDEX

Alvarez, Al: stretching one's ear toward the person speaking 81
anxiety, acute versus chronic *see* Bowen
Apprey, Maurice: *transgenerational haunting* 108
Arendt, Hannah: *The Human Condition* 5; philosophy as thinking within and for one's community 22; the "shame at being a human" 28
aspiration: concept as developed within TA community 93 (*see also* Berne; Clarkson); as dependent on culture 107; emerging in relation to the human condition 5; a fragile arc 93; as the fulfillment of the genome across its achievable life span 147; as moving group purposefully through disorderly processes 78; as necessary for learning organizations and schools that learn 117 (*see also* Senge & colleagues); as an outgrowth of researching 77; as serving the good of the individual and of the group 93

bacteria: as capable of learning 101, 103; as colonies representing the evolved capacity to cooperate 103, 111; learning processes 101–102; as a social species 102
Barrow, Giles: acting swiftly on instinct that something is amiss, not waiting or wondering 83; culture as husbandry or farming 41, 107; desire of individuals and groups to develop and flourish 77; link between education and human gifts for responsive cultivating 71; somatic and environmental impacts of educational encounters 71

Barrow, Giles & Newton, Trudi: education involving an act of faith 71; educational transactional analysis 71
Berg, Jeremy M. & colleagues (Tymoczko and Stryer) 101
Berne, Eric: Adult ego state 30, 35, 55, 59, 64, 67, 80, 105; aspiration 68, 73, 93; bodily sensing as the first step in an intuitive process 36; capacities for awareness, spontaneity and intimacy 17, 75, 106; the contract, psychological 76; contracting as bilateral process 130; on culture, early writings 40–41; on curing 18; *differentiation* as distinctions among classes of objects and among individuals in a group imago 106 (*see also* Bowen, on differentiation of self); dismissal of culture as affecting script 40–41 (*see also* Mazzetti); the existential motto or dynamic slogan 91; existential positions, Kleinian influence upon 33 (*see also* Harley); no hope for human race as a collective but hope for individual members 76; hospital admissions correlated with being told to drop dead 56; human tendency to take and to destroy anything getting in the way 106; humans continually and quickly making subtle judgments about one another 29; intuition defined 137; intuition related to cognitive processes in lower animals 28–29; listening in the Adult and Child ego states 35, 80; making decisions without conscious thought 23; the Martian viewpoint 109; mortido 44; multiple modalities

of relating 50; nervous tissue 24; *physis* defined 95; the problem of a human being 73; *protocol* 13, 14, 44, 130 (*see also* Cornell & Landaiche); reality, understanding it and changing our corresponding images 35, 98, 109; therapeutic operations 71; "toward hard therapy and crispness" 42

Berne, Eric, on group leadership: as attending to many forms of communication, including the nonverbal 118; evaluating leadership based on outcomes 127; a leader composes her or his mind for the work that lies ahead and starts anew with a fresh frame of mind 87, 109–110, 123; leader's duty to observe every movement of every muscle during every second 141; leader's focus on her or his personal development 51–52; on learning something new every week 87, 109–110, 123, 136; physiological signs to be observed 81

Berne, Eric, on groups: on Bion's practice of observation in groups 11–12; the combat group 105–106; contracting for groups 129; cross-cultural factors in group therapy 39; culture in groups defined 40–41, 107; early writings on groups 38–39; extrusion from group as expulsion, excommunication, discharge, or rejection 60; forming groups to prevent biologic, psychological, and moral deterioration 46, 97; group canon's centrality 40, 41; group culture's preservation taking precedence over group activity 42, 107, 114; group culture's preservation tantamount to fighting for survival 107; group extrudes agitator perceived as too great a threat 60; group's first task, to ensure its survival 105, 106; groups of all sorts and sizes, from nations to psychotherapy groups 38; groups as frightening and perilous to the unprepared 76, 77; model of the group as a whole in dynamic relations to the model of the individual human personality 106; social aggregation with external boundary 38; survival of group requires preservation of organizational structure 114; on suspending the group's primary work 105–106; transactional analysis as a group therapy, derived from the group situation itself 38, 75; types of groups that Berne led, observed, and studied for *The Structure and Dynamics of Organizations and Groups* 141

Berne, Eric, on staff-patient staff conferences: the abolition of professional categories 42; acknowledging participants as individuals, not as roles 106–107; intended to stimulate thinking and the organization of thoughts 49, 106–107; interdisciplinary considerations 72

Berne, Eric (writings): *Games People Play* 24, 75, 76, 77, 106; *Intuition and Ego States* 28, 36; *A Layman's Guide to Psychiatry and Psychoanalysis* 35, 98; *The Mind in Action* 23, 44, 73, 95, 106; *Principles of Group Treatment* 18, 38, 42, 44, 50, 51–52, 71, 76, 81, 87, 109, 118, 123, 129, 130, 136; "Staff-Patient Staff Conferences" 39, 42, 49, 70, 72, 106–107, 127; *The Structure and Dynamics of Organizations and Groups* 11–12, 38, 40, 42, 44, 46, 60, 97, 105, 106, 107, 114, 129, 131, 141; *Transactional Analysis in Psychotherapy* 38, 75; *What Do You Say after You Say Hello?* 35, 40–41, 44, 56, 68, 73, 80, 91, 93, 109, 141

Berne's group legacy 38–40

Bion, Wilfred R.: *alpha function* 15–16, 105; *beta elements* 14, 15–16, 105; bodily drive to think 98–99; bodily drives to love, to hate, and to know 22, 98–99; containing function 12–17, 18, 19, 21, 22, 105, 33–34, 49, 51, 66, 67, 74, 83, 89, 105, 133, 136; disturbed patient aware of something the analyst may not want to see 16; hatred of emotions but a short step to hatred of life itself 35; individual having to live in a body and to put up with the mind living in it 24; learning from experience 137; philosophic doubt as the psycho-analyst's tool 109; processing at a bodily, felt level 11; risk of turning ourselves into receptors 11; sequence of communication, reception, reflection, and considered

INDEX

response 13, 14–17; on thinking 114; thinking called into existence to cope with thoughts 98; unbearable tensions between loving and hating, between self and other, between knowing and not knowing 18

Bion, Wilfred R., on basic assumptions: as corrupting generative leadership 128; as an impasse in group's or system's capacity to process or integrate the data of experience 125; as instantaneous, inevitable, and instinctive, requiring no training, experience, or mental development 104–105; as leading to fight, flight, or dependency 48, 75, 100, 128; on less rational use of verbal communication in basic assumption groups 118; as potentially destructive 103, 108; as predominating over productive functioning of *W*[ork] group 106

Bion, Wilfred R., on groups: difficulty of understanding a leader who neither fights nor runs away 128; group is following a lead unless that lead is actively repudiated 127–128; group members' hopes of satisfaction paired with a sense of frustration with the group 46; group's emotional situation interfering with clearheadedness 49; group's high opinion of designated leader a sign of basic assumptions 128 (*see also* Bion, on basic assumptions); groups as essential to fulfillment of human mental life as groups are essential to economics and war 75, 97; individual's inevitable membership in a group even when behaving as if not belonging to any group 47; limited usefulness of making a group member or leader an established or discredited god 127; verbal exchange a function of the work group 118; *W* group as work group 105, 106

Bion, Wilfred R. (writings): "Attacks on Linking" 35; "Attention and Interpretation" 12; *Bion in New York and São Paulo* (1980) 11, 16; "Elements of Psycho-Analysis" 12; *Experiences in Groups and Other Papers* 46, 47, 48, 49, 75, 97, 100, 104–105, 105, 118, 127, 128; "Learning from Experience" 12, 99, 137; *The Tavistock Seminars* 24; "A Theory of Thinking" 98, 114

bodily capacity, development for professional work 74–75

bodymind 5–6, 24–25; befriending the bodymind 35; bodymind integration 24, 26; bodymind tension 24–25; receptive bodymind 36; reciprocal regulation 30, 31, 34, 61; splitting off 14; thinking with the full bodymind 37; *see also* shared bodymind; unitary bodymind

Bollas, Christopher: *free association* 138; learning as a desired transformational experience 31

Bowen, Murray: anxiety, acute versus chronic 100, 104 (*see also* leadership/ followership roles; anxiety); anxiety as system-level movement and responsiveness 29, 49, 96, 103; automatic forces governing protoplasmic life 95; *chronic anxiety* 100, 104, 105, 125, 131, 136, 144, 146; human *emotional system* 29, 95, 97, 102, 103–104, 112, 133; human *intellectual system* 29, 102, 103–104, 105; lawful order between cell and psyche 96, 101; overfunctioning 45, 53, 89, 90; the part one plays in a systemic problem or symptom, acknowledging and addressing 132, 139; research methodology 111, 112, 125, 126, 132–133, 137; societal regression 125; a theory is never absolute or complete 110; two opposing life forces, individuality and togetherness 104; undifferentiation 29

Bowen, Murray, on *differentiation of self* 29, 104, 105, 106; concept borrowed from cell differentiation in biology 104; process of differentiating a self 112, 125, 126, 132–133; *see also* Berne

Bowen, Murray, on family: chronic anxiety and emergence of severe symptoms 104, 131; as evolved over time to be an emotionally interlinked and responsive system 29, 96, 103; family leader defined 129; family's chronically anxious process 125; from family unity comes individual diversity 97; individual behavior understood only in the context of the

INDEX

family 3, 29, 96, 128–129; multigenerational transmission process 107, 134, 146; as a natural, living system 29, 96; with a psychotic member 96, 131; as a resourceful organism 96, 131; tension between togetherness and separateness, between groupish, less-differentiated behavior and individual initiative that uses the mind to override automatic impulses 29, 128–129; undifferentiated family ego mass 129

Bowen, Murray (writings): "Diversity from Unity" 97; "Epilogue: An Odyssey toward Science" 129, 132; *Family Therapy in Clinical Practice* 29, 95, 96, 103, 104, 107, 112, 125, 128, 131, 132; "A Psychological Formulation of Schizophrenia" 96

Brooks, David: the knowledge transmitted in an emotional education 75

Bushe, Gervais & Marshak, Robert: *dialogic organization development* (OD) contrasted with classic diagnostic organizational development 119; good dialogue not sufficient for transformational change 120; methods of dialogic OD 120–121; shared characteristics of dialogic OD 119–120; *see also* Shaw; Stacey

capacity: as a bodily ability to contain or hold 11, 66, 74, 133; development of 74–75; for being oriented to reality and living with the facts 17, 37; for human speech 122; for leadership 87, 98; for neurophysiological integration 73, 100, 124; to sit silently with restraint 79, 136; for separating from the group's compelling togetherness 129; for suffering the challenges of life 105

Caper, Robert 17; analytic containment not about feeling good but about thinking and feeling what is true 19; containment helps patient bear not just current state of mind but also future ones 18; fate of an analysis determined not by analyst's interventions but by dynamics of patient's unconscious 18

Chinnock, Keith: integration of learning models for the supervisory process 71; *see also* Newton

Clarkson, Petruska: aspiration, conceptualization of 73, 93

communication: giving priority to the unworded message 36; overt message and the more emotional layer 36; *see also* nonverbal communication

compassion 28, 36, 37, 96, 126

containing, an example of weaning a child from the breast 12–13; *see also* object relations theory

contracting: for learning outcomes 76–77; in transactional analysis tradition 129; *see also* Lee

Cornell, William: on aspiration versus adaptation 93; establishing communities that cultivate thinking 114; fostering freedom for play, imagination, and uncertainty in professional learning environments 114

Cornell, William F. & Hargaden, Helena: relational aspects of transactional analysis 24

Cornell, William F. & Landaiche, N. Michel: on protocol 44; unavoidable intimacy of the working therapeutic or professional couple 61

Crespelle, Isabelle: training psychotherapists using a model of active observation and reflection on the group process 39, 70

culture: as by-products, climate and action 41–42; as a paradigm 108–109 (*see also* Kuhn); Schein's definition of 108; as sets of tools or schemas, environments or atmospheres, and the intentional shaping of life 41, 107; transmission of 107–108

culture of transactional analysis community *see* transactional analysis, community culture

Dalal, Farhad: growth model critiqued 91; power dynamics of racism 90; power relations 89; tyrannies of internalism and individualism 88

INDEX

Damasio, Antonio: culture as developing in a biological and social context 41–42

Darwin, Charles: emotional intensity confuses mental powers 61; observing "violent action under extreme suffering…to escape from the cause of suffering" 58–59; writings on emotion prefigure contemporary links among social pain, shame, humiliation, embarrassment and observable physiological signs 58; *see also* social pain response

de Graaf, Anne & Levy, Joost: maintaining a learning environment that is as safe and as disturbing as possible 84

Deaconu, Diana: searching for group in self, in other, and in community 39

defining a self in one's system: engaged researching practices 132; establishing one-to-one contact with other individuals in the system 133; making contact with the system 133; managing oneself in the system 133; observing oneself in the system 133

Dervil, Yolle-Guida: applying the ideas and practices of Thomas Schur 118

Dewey, John: body-mind 24–25

dialogic organization development 119–121; *see also* Bushe & Marshak; Shaw; Stacey

Diamond, Jared 144–145

Dickinson, Emily: "a Panther in the Glove" of our skin 23

distress in clients as a profound disruption in human learning and integrating 72

Douglas, Mary: *How Institutions Think* 41; institutional ways of coordinating essential activities 49, 114

Downing, George: use of video to help troubled mothers see interactions with their children 19–20; *see also* object relations theory

Eigen, Michael: our ability to process experience falls short of the experiences we must process 14; reciprocal bodily regulation with patient 34; religions, psychotherapies, art, and literature as frames of reference for processing the unbearable 16

Eisenberger, Naomi & colleagues: chronic physical pain and maturation 57–58; *pain matrix* 56; social pain research 56

Eliot, George: "the mysterious complexity of our life" 124

emotional learning in unavoidable groups 75–76

encountering group life, an intersection of frameworks from Berne, Bion, and Bowen 6, 149

engaged researching 7, 18–20, 21, 38, 136; after-effects of observing 20–21; containment delivering a therapeutic effect 21, 22; a contribution to thinking for and within one's community 22; difficulty of living this engaged way of working 21–22; researching human systems injects thoughtfulness, curiosity, objectivity 22; observing a training group 9–11; participative use of the researcher's full being 20

English, Fanita: "The Three-Cornered Contract" 129

English, Fanita, on shame: de-shaming antidote treatment 67–68; feeling of shame as a psychosomatic reaction 58; identifying feelings and behaviors associated with shame, then confronting with Adult examination 67; physiological and behavioral signs of shame 65; shaming as an evolutionary development for maintaining standards in human societies 60

Ernst, Franklin H., Jr.: intensity of shame, of feeling rejected (or inferior), and flashes of anger, all linked to life positions 58; *life positions* 58

Erskine, Richard: overview of literature on shame 58

Evans, Kenneth: the impulse to solve a problem rather than learn to live with it 66

facilitative practices for group learning: asking questions to foster clarity amidst frustration and confusion 51,

INDEX

54; forming hypotheses for guidance 125–126; holding space for everyone 139–140; learning continually 135–136; learning experientially 136–137; listening to the content and its bodily effect 140; as a multigenerational transmission process 134–135, 142; as representing the best thinking of the elder generation 135; researching further 141–142; sharing hypotheses 140–141; speaking for self 139; thinking and speaking freely 137–139; *see also* practices and principles of group work

facilitator's role: assuming group members have capacities to learn, areas of difficulty, and priorities for growth 76; attending to learning aspirations and impasses 76; attending to needs and troubles of group members and of the group as a whole 86; intervening at first sign of possible harm 83; looking for and making trouble 69, 84–85, 85; posing particular questions to orient amidst the process 77–78; safeguarding opportunities for group members to talk of their experiences 80; scanning for trouble 80; speaking when data of experience have organized themselves sufficiently for description 83; suffering terror 85; watching for and supporting whatever learning emerges 77; *see also* leadership in groups

family systems *see* Bowen

Fanon, Franz: final prayer 37

Fast, Irene: body-centered mind 25

followership, benefits of 132; *see also* leadership/followership

Fowlie, Heather & Sills, Charlotte: relational transactional analysis 71

Fox, Leslie Ann & Baker, Katherine Gratwick: leadership as a reciprocal process 129

free association see Bollas

free speech, integrating potential of 79–80

Freire, Paolo: the teacher is taught in dialogue with the students 86, 135

Freud, Sigmund: the bodily in the mind *see* Fast

Furness, John 26

Gershon, Michael 26

Gilligan, James: prisoners' histories show that efforts to escape affective intensity and suffering underlie many extreme acts of interpersonal violence 59; *see also* social pain response

Goldberg, Carl: curiosity about the experience of the one who is shaming 67

Grant, Jan: approaches to transactional analysis training correlated with the principles of effective adult education 70

Gray, Leila 101

group learning, importance of freedom for authentic expression 121

group life, continual tension between creativity and deadness 120

group nonlearning and learning assessed through individual and group use of language in dialogue and in debate 119

group tools: a boundary, confidentiality, focus for work, time and space structures, ethos of not harming, freedom to speak, experiential exercises, capture of group learning 142

group's body as a mansion with many rooms 121

groups: able to think 48–49; as brainless yet capable of learning 99–100, 103; as clustered creatures 94; as a cognitive-like process capable of integrating stimuli from environment and from within group itself 100; comprised of interacting, cross-regulating human bodies 95–96; as distributed systems of processors 100; embodying traits common among different forms of life, i.e., metabolism, reproduction, responsiveness 96, 102–103; *groups that learn (and don't)* 8, 95–112; as unavoidable and necessary 54, 72, 75, 136; learning capabilities 99, 100; as learning systems 99–103; as living entities or systems 95–97, 99; problems presented by limited time and resources, disorganization, conflict, banal groupthink, peer pressures, boredom, passivity,

scapegoating 75; as unavoidable 54, 72, 75, 136
groups that don't learn 103–107; problematic dimensions identified by Berne, Bion, and Bowen, i.e., lethal games, destructive basic assumptions, and unregulated emotional processes 103
groupthink 48, 75, 104
gut responses, enteric system neurons 26

Hacking, Ian: on Kuhn's conception of progress in science 147
Hanna, Nick viii; good group membership as good group leadership 129
Harari, Yuval: the agricultural revolution as history's biggest fraud 144; impact of key technological decisions on human wellbeing 145
Hargaden, Helena & Fenton, Brian: reciprocal influence and mutual regulation 34
Hargaden, Helena & Sills, Charlotte: relational transactional analysis 71; therapy groups as living, breathing organisms 48
Harley, Ki: Kleinian influence on Berne's development of the existential positions 33
hating in groups 46–47
Hawkes, Laurie: on use of tango in TA 39, 115
Hinshelwood, Robert D.: humble gloom of practitioners who know that nearly everything remains uncertain and paradoxical 76; the individual's problem of securing personal identity in the group 50; personal despair correlated with organizational demoralization 53
hoping within reason 147–148
Hopkin, Michael 101–102
human condition 5, 17–18, 19, 43, 44, 56, 88, 90, 93, 124; *see also* Arendt
human intelligence or cleverness 144–146
human toolmaking and symbolizing 146
Hyams, Hanna: "The deeper the shame, the more violent the hatred" 58 (*see also* social pain response); individuating oneself from the problem 67

impediments to growth 72–74
individual and the collective, in tension and in complementarity 88–89, 93, 103
individual in the group 88, 90, 92, 93
individuality as understood in its encompassing human context 96–97, 100, 104, 106–107; *see also* Bowen
an integration of group work using the theories and practices of Berne, Bion, and Bowen 6, 149
interdependence of self-care and care of the group 126–127
interdisciplinary tradition in transactional analysis 43, 72; *see also* Landaiche
interpreting as a function of productive containing 16–17
ITAA conference on shame, collected papers edited by Norman L. James 58

Jackson, Philip & colleagues (Meltzoff and Decety): neural processes involved in empathy 62; *see also* Singer & colleagues
Johnson, Mark: mind-body interdependence 25; studying our embodiment 6
Johnson, Mark & Rohrer, Tim: cognition as constituted by social interactions and relations 102; from single-celled animals to the highest cognitive achievements of humans 96; thinking as a form of bodily action in the world 102
Joseph, Marina Rajan: Berne's "therapeutic operations" as relevant interventions for teaching 71

Kolb, David A.: challenge of lifelong learning as the challenge of integrative development 74, 98; dialectic tension between experience and analysis 137; experiential learning 78, 136
Kreyenberg, Jutta: organizations as living systems 48
Kuhn, Thomas S.: characteristics of a scientific community 114–115; community structure of science, importance of 115, 148; the limits of progress 147; the need to study the community structure of every field of

INDEX

practice 122, 148; normal science 99, 110, 134–135; on progress in the sciences 147; scientific community consisting of individuals who share a paradigm 114–115; scientific knowledge, like language, a common property of a group 122; *The Structure of Scientific Revolutions* 99, 108, 110, 114–115, 122, 134–135, 146–147, 148

Kuhn, Thomas S., on paradigms: changes in paradigms as not leading scientists closer to the truth 147; as constitutive of nature as well as of science 108; as dogmas that no longer represent actual research findings 135; as guiding research through direct modeling and abstracted rules 108; *paradigm*, the shared assumptions of a scientific community 108, 115; as shaping and reinforcing normative behaviors and perspectives 108; transmission of scientific paradigms 134–135

Lakoff, George & Johnson, Mark 26

Landaiche, N. Michel: interdisciplinary tradition in transactional analysis 43, 72; learning, maturing, and integrating defined as intention to research and then to revise initial theory 111; on succession and reproducing life 134; on researching and presenting one's family system 126; review of Barrow & Newton book on educational transactional analysis 71–72, 72, 73; on skepticism 109; turbulence as requisite for cross-generational learning, development, and maturation 122

language: its role in the life of our groups 118–121; nonverbal transmissions 80–81, 118; *see also* Schur

Lawrence, W. Gordon & Armstrong, David: importance of experiencing psychotic edge in order to work with destructiveness and creativity in organizational life 74

leadership in groups: capacity developed concurrently with effort to facilitate development for the group and its participants 87; designated versus actual 115, 127–128; as an emergent property of groups 127; emerging from anywhere in the group 127–128, 132; emerging from willingness also to be a member still learning with others 87 (*see also* leadership/followership); examining one's own membership patterns in groups 3, 38, 53, 54, 86, 129; failure of leadership resulting in piercing shame 28, 85; on incapacity to contain group's experience in order to facilitate growth 83; as managing one's personal problems with group membership 86–87; as overfunctioning vis-à-vis sibling position 131–132; a position of continually not knowing 85–86; *see also* facilitator's role

leadership/followership 127–132; contractual nature of, implicit and explicit 129–130; intuitive emergence vs. intentionally defined roles 128, 130–131, 132; a lead is simultaneously given and followed 129; manifestations differ based on degree of separateness and togetherness in group 131; necessary interdependency and flow 129; reciprocal nature of 129; roles vis-à-vis tasks, anxieties, and learning 129–131

learning: allowing for evolutionary change and survival across generations 103; to apprehend or grasp 97; as distinct from trauma 99, 107–108; equated with scientific method 109–110, 111; guiding hypotheses 124–125; as integration of experiences 8, 72, 98, 111–112, 124–125, 125–126, 133, 136, 147; as the intention to research 111; Latin root *lira* indicating furrow or track 97; as making meaning 98; as necessary for ongoing life in a changing environment 103; as the neurophysiological integration of the data of experience 98; as a process common to all life forms 101–102, 102–103; *see also* scientific method, in relation to learning

learning and hating in groups 7, 38, 47–47

learning community: characteristics of 114–115; definition of 113; as an ecosystem 114; embodying ethical stance of concern for individual members, group as a whole, and future generations 114; examples of 113, 115–116, 117; Kuhn's characteristics of the scientific community 108, 114–115, 122, 134–135; learning collectively about the process of human learning 113–114; scientific community as exemplar 5, 113; thinking collectively as differentiated individuals within a cooperative human system 113

learning continually, importance for the teacher, trainer, group leader, and professional 86, 135–136

learning facilitation, some orienting questions 5, 51, 54, 77–78, 123, 126; *see also* facilitative practices for group learning; facilitator's role

the learning organization, fostering continuous learning of members in order to transform organization and its capabilities 117; *see also* Senge & colleagues

learning, pain-cued: to sustain group's coordinated efforts and activities, and to reinforce and coordinate swarming, shoaling, flocking, and herding 60

Lee, Adrienne: *process contract* 77

Lerkkanen, Marja-Kristiina & Temple, Susannah: effects of transactional analysis content and processes on personal and professional development of teachers 70

lesson plan, abandonment of 78

Lewin, Kurt: *action research* 21, 123

Lewin, Melanie: individual's separation distress followed by violence of shunning significant others 61

Lewis, Thomas & colleagues (Amini and Lannon): limbic system mutually cued 29–30

limbic system 29

listening with our bodies 50, 80–82

looking for trouble in groups 8, 69, 84–85

MacLean, Paul: the human *triune brain* 57; sense of separation a painful mammalian condition 57

Manor, Oded: "Transactional Analysis, Object Relations, and the Systems Approach: Finding the Counterparts" 6

Maquet, Jean: difficulty of pinning down the psychological contract in professional relationships 76

Marcum, James: older paradigm is like a fossil 143; world as a niche to which science helps a community adapt 112

maturational impasses, a case example of teaching 31–33

maturing as a community effort 8, 91–93

maturing for individual of vital interest to the group 92

Mazzetti, Marco: arguing against Berne's late-life dismissal of culture as impacting scripts 41

Mellor, Ken: communicating acceptance as the means of providing help 66

membership in groups 3–4, 38, 45, 46, 53, 54; as a prerequisite for productive leadership 53; *see also* leadership in groups

Milosz, Czeslaw: admitting, amidst clamor, to not understanding 51; divesting ourselves of shame in order to speak 49; shame about helplessness and ignorance 49, 138; speaking of things not yet understood 138

Minnich, Elizabeth K.: teaching thinking to young adults, importance of 114; thinking within and for one's community 22

Morrison, Toni: on the pieces of oneself being gathered and given back by the other 35

Napper, Rosemary: co-editor of journal issue dedicated to training in transactional analysis 70

Newton, Trudi: basic human process of generating hypotheses 78; consensual imagoes 44; editor and co-editor of journal issues and books dedicated to field of education and training in transactional analysis 70; gift economy or culture of transactional analysis community 43, 89; integration of learning models for the supervisory process 71 (*see also* Chinnock); nature, necessity, and ethics of risk ineducational settings and processes 71; passing on the TA "OK-OK" culture 41; real learning as a therapeutic process 71; transactional

analysis beginning as a group therapy 75
nonlearning, as an anxious approach to or attitude about science 111
nonverbal communication via body postures, qualities of eye contact, breathing patterns, restlessness, calm, agitation, fear, boredom, withdrawal, and excitement 75; *see also* Berne, on group leadership
Novellino, Michele: Berne's concept of the Demon re-examined 44; the mind's primary need for interpersonal regulation 34

object relations theory: analyst-client relationship as analogous to mother-infant pair 34; interdependent maturational processes 33; mind-body split 26, 33; mother-child dyadic interaction 12, 13; nonconscious communication via the shared bodymind 33; outgrowth of classical Freudian precepts 33; psychotic processes operative for all humans to varying degrees 33; Samuels, Charelle, mentor viii–ix; *see also* containing; Downing
Ogden, Thomas: on Bion's *container-contained* 33–34; thinking as an ongoing, interdependent process 30; thinking requiring at least two people 30; on Winnicott's *holding* 33–34
organizational engagement, example of 27–28
overfunctioning *see* Bowen

pain matrix 56; activation and nonconscious learning 56–57; activation in relation to empathy 62; *see also* Eisenberger & colleagues; Jackson & colleagues; Singer & colleagues
pain-cued learning can sustain group's coordinated efforts and activities 60
pain-inducing behaviors involved in conforming and social control 60
Panksepp, Jaak: intense pain alleviated by release of opioids in the brain 57
parallel process 20
personal group psychology, example of 44–45

personal histories with groups 2–3
Petriglieri, Gianpiero: identifying what one wants to learn tomorrow as a guide to further researching 22; respected marginality of the transactional analysis community 44
Petriglieri, Gianpiero & Wood, Jack: relinquishing the counterproductive aim of controlling the group 52
phantasy, Kleinian 13–14; *see also* protocol
phenomenology viii, 4, 19, 25, 110
physiological interrelatedness, an organizational example 27–28
power relations and accountability 89–90
practices and principles of group work 123–142; practices as methods or particular activities 134, 135; principles as guidelines and values for practice 123–124, 135; *see also* facilitative practices for group learning
Pratt, Karen & Mbaligontsi, Mandisa: using TA concepts for grassroots community care workers in South Africa 70
principles of group work *see* practices and principles of group work
processing lived experience 15–16; *see also* Bion, alpha function
professional capacity: as increased understanding of human functioning, ability to sit with the fullness of human experience, and awareness of how psychophysiological maturation is facilitated 76; as leadership 87; *see also* leadership in groups
professional susceptibility to social pain 61–62; countertransference 62; dependence on clients or students for income, job satisfaction, and/or referrals 62; empathy activates pain matrix 62 (*see also* Jackson & colleagues; Singer & colleagues); inherited or learned sensitivities to rejection 62; necessity of gathering nonconscious, highly emotional information 62
progress and mortality 146–147
projection 14–15
protocol 13–14, 44, 130; *see also* Cornell & Landaiche; phantasy

INDEX

psychoanalytic terminology as utilized within transactional analysis community 20

Ramanujan, Krishna 101
receiving signals and other communications from the other 11–12
rejection: social function of 57, 59–60; violence of rejecting as a paradoxical outcome of shame and social pain 60–61
researching *see* engaged researching
Rogers, Carl: potency of groups, both constructive and harmful 45; professional risk of launching into a therapeutic relationship and possibly failing 85
role lock *see* Sills

Samuels, Andrew: engaging the world as a broken, fractured, and stunted individual 92; the individual's role in the collective 88; the limits of individual responsibility 90
Sardello, Robert: involvement in world to allow reality to reveal itself 110; on the phenomenological perspective of respectful openness to existence 19, 110
Saru, P. K.: training as a process whereby trainer and participants mutually evolve, learn, and integrate 86
Schanuel, Marilee: caution against emerging insularity in transactional analysis community 44
Schein, Edgar, on culture: definition of 108; as a product of joint learning leading to shared assumptions 108
Schmid, Peter & O'Hara, Maureen: moments in group life when intractable complexities are reconciled 50; reading the signs expressed in the non-linguistic patterns of group life 81
Schore, Allan: attachment as dyadic regulation of emotion 30; principles of regulation theory apply to mother-infant and clinician-patient relationships 34; *see also* object relations theory
Schur, Thomas: Bowen theory applied to supervisory role 118; *changing self in the conversation* 118; conversation between supervisor and supervisee as part of therapeutic system with supervisee and client 118; differentiation of self as seen in person's use of language for reflection and conversation 119; experimenting in conversation to discover what is most effective for growth within a system 118; function of language and conversation in natural human systems 118; language as neurological coordination more than as delivery of information 119
science: as equated sometimes with absolute knowledge and rigidity, an anxious manifestation 111; as ethical quest for truth 110
scientific method, in relation to learning 109–112; *see also* learning
self-analysis 24
Senge, Peter & colleagues: disciplines that support organizational and community learning 117; the *learning organization* 117; schools that learn, characteristics from a systemic perspective 117
separation and rejection, differences and commonalities in terms of threatening interpersonal ruptures 57
separation distress 57–58
shame response, physiological 58; social pain correlated with embarrassment, blushing, elevated heart rate, collapsed body posture, averted gaze, and blunted cognitive capacities 58
shame response to rejection ix, 56; *see also* Thomas
shaming and bullying in schools as saturating entire school atmosphere and negatively affecting learning for all 73
shared bodymind 7, 28–31, 32, 33, 34, 61, 95–96; *see also* bodymind
Shaw, Patricia: conversing as organizing 79, 119; *dialogic organization development* 119; human conversing as continually recreating and reevaluating the organization and its function 119; human organizations as the outcome of human conversing 119; organizations characterized less by formal structures and more by the way language is used to achieve (or

174

INDEX

not) organizational purposes 119; *see also* Bushe & Marshak; Stacey
Sills, Charlotte: role lock 48, 52
Singer, Tania & colleagues: empathy for pain involves affective rather than sensory components of pain 62; *see also* Jackson & colleagues
skepticism: absence correlated with harming 28; as corporal 36; as important attitude 109; in relation to compassion 36, 37
social pain, challenges in practice 65–68; embodying acceptance 66; gathering courage 67–68; identifying the signs 65–66; shifting to objectivity 67; *see also* professional susceptibility to social pain
social pain, managed and reduced by: maturing on the part of clients, students, and professionals alike 57–58; attaining Adult ego state 64–65, 67
social pain, reciprocally activated 60–61; everyone in system can become implicated 61; group's rejecting or controlling behaviors as a result of social pain 61
social pain resolution as shift from violence to insight 67
social pain response 7; behaviors associated with 57; correlation between violence of feeling and violence of action taken to eliminate it 58–59 (*see also* Darwin; Gilligan; Hyams); distinct from physical pain 56; intensity directly correlated with significance of the persons rejecting 57; interrelated aspects of rapid corrective action, separation distress, shame response, and correlation between affective and interpersonal violence 56; norm-shaping role 60; relieved by use of right ventral prefrontal cortex (RVPC) to think more objectively 56; a threatening interpersonal rupture 57, 63; transactional model of 59–61; *see also* Eisenberger & colleagues
social pain transaction as a relayed visceral intention and subsequent reception with interpretation 62–63
social pain transaction, analysis of 62–65; analyzing the transaction 64–65; bracketing other ego state content 63–64; receiving and interpreting the intention 63; relaying the intention 63; role of the Adult ego state 64
Solomon, Carol: on Berne's method of listening in groups 83–84
speaking authentically, correlated with researching self-in-system (family, group, organization, or community) 119
spontaneity, awareness, and intimacy *see* Berne
Stacey, Ralph: anxiety as inevitable companion of change and creativity, along with destructive interruptions in communication 120; communicative interaction producing collaboration and novelty along with sterile repetition, disruption, and destruction 50, 120; shifts in patterns of communication indicate shifts in power relations and in patterns of inclusion and exclusion 120; *see also* Bushe & Marshak; Shaw
Stacey, Ralph & Griffin, Douglas: understanding organizations through participation 47
Stern, Daniel: *vitality affects* 14
studying ourselves in collective life 4; basic research questions 54; Bucharest study group members ix; as a living science supporting integration more than anxious, harmful nonlearning 112
Sullivan, Harry S.: unique individuality a delusion 31

tango community, characteristics and learning processes 115–116; *see also* Hawkes
Tavazoie, Saeed & colleagues 101–102
Tavistock approach to group work, grounded in views of Bion and others 3; Bonds-White, Frances, mentor viii
teaching in groups as a potentially disturbing process 69–70, 84–86
thinking as the organization of conscious and nonconscious sensations from multiple forms of experience 31; *see also* Bion
Thomas, Herbert E., on shame: avoidance of painful shame leading to

INDEX

eccentric, paranoid, or psychotic withdrawal 59; intensity of shame directly correlated with significance of person or persons doing (or even witnessing) the rejecting 57; intensity of shame inversely correlated with psychological maturity or capacity for psychological separation from significant others 57; objectivity as crucial for resolving a painful shame reaction 67; objectivity as imagining the inner state of the rejector 67; *The Shame Response to Rejection* ix, 56, 57, 59, 67

Thompson, Erik: on importance of finding courage for leadership 132

transactional analysis, community culture 41–44; belief in capacities for growth and healing 43; climate of vigorous exchange 52; freedom to be curious about groups and to use them in myriad ways 40; gift culture with sense of social responsibility 41, 43 (*see also* Newton); independent and sometimes rebellious turns of mind 42; interdisciplinarity 43 (*see also* Landaiche); offering strokes while holding "OK-OK" position 43; over-emphasis on thinking and rationality 44; permission to provoke, play, disagree, and make trouble 42; predominant pursuit and nurturing of development 52; a shared language that remains open to other frames of reference 42–43; splitting and divisiveness among fields 44; structuring time and work in a manner that requires group participation 42

transactional analysis, educational field: consideration of educational metaprocess separate from technique or content 70–71; guidance from transactional analysis for learning to teach 70–72; *see also* transactional analysis literature, on educational field

transactional analysis, international community ix, 3, 40, 89

transactional analysis, relational 71; *see also* Cornell & Hargaden; Fowlie & Sills; Hargaden & Sills

Transactional Analysis Journal: Robin Fryer, longtime managing editor and mentor ix

Transactional Analysis Journal, theme issues: cultural scripts 41; education and training 70; group work 39; organizational transactional analysis 39; shame 58 (*see also* ITAA conference on shame); transgenerational trauma 107–108

transactional analysis literature, on culture 41; *see also* Berne on culture, early writings

transactional analysis literature, on educational field 70–72

transactional analysis literature, on groups: applications with special populations 39; broader spheres of community and politics 39; comparison of different theoretical and practice approaches 40; group dynamics across differing contexts 40; groups for counseling and psychotherapy 39; interplay between institutional factors and groups 40; organizational systems and consulting 39; personal perspectives 40; training and education 39; *see also* Berne's group legacy

transactional analysis literature, on organizational theory and practice 39; *see also* Berne (writings)

transgenerational process: concern for the next generation 42, 69, 96, 100, 114, 134, 143–144; inheriting much from the bodies and minds of those who came before 100, 107, 135, 143, 149; *see also* Bowen, on family

trauma as interfering with integration of experience 82, 99

trouble with having a human mind with its human body 23, 24, 73

troubles that bring people for help include thinking what is not necessarily real and an incapacity for bodily affect 23, 73

Tudor, Keith: group facilitator also lives the uncertainty of discovery 80; teaching in the manner of TA 42, 70; training structure of transactional analysis as unique 70; all transactional analysis trainers as being in the field of education, no matter their areas of primary practice 72

INDEX

turbulence, at tolerable levels, as requisite for cross-generational learning, development, and maturation 122; *see also* de Graaf & Levy

undifferentiated ego mass 53; *see also* Bowen
unitary bodymind 24–26

van Beekum, Servaas: inviting rather than soothing trouble 84; on not taking clients to deeper levels than one has the courage to go oneself 86–87; relational consultant, importance of sitting with restraint 79
van Beekum, Servaas & Laverty, Kathy: emergent experiences that are not exclusively the product of the individuals involved 48–49

van Poelje, Sari: optimal learning opportunities for developing organizational leaders 70
violence, affective and interpersonal 58–59; *see also* Gilligan

Western Pennsylvania Family Center: Paulina McCullough, James B. Smith, and Walter Smith, mentors viii, 3
Wheatley, Margaret: groups and communities as living systems 99
Williams, Kipling D. & colleagues (Forgas, von Hippel, and Zadro): pain matrix evolutionarily co-opted to signal social pain given importance of group inclusion for human survival 57
Winnicott, D. W.: environment-individual set-up 30; facilitating maturation 8; holding 33, 34; no such thing as a baby 30

Milton Keynes UK
Ingram Content Group UK Ltd.
UKHW020749221123
433026UK00014B/97